Electrodeposition of Coatings

Electrodeposition of Coatings

A symposium sponsored by
the Division of Organic
Coatings and Plastics
Chemistry at the 161st
Meeting of the American
Chemical Society, Los Angeles,
Calif., March 31 - April 1, 1971.

George E. F. Brewer,
Symposium Chairman

ADVANCES IN CHEMISTRY SERIES **119**

AMERICAN CHEMICAL SOCIETY
WASHINGTON, D. C.   1973

ADCSAJ 119 1-243 (1973)

Copyright © 1973

American Chemical Society

All Rights Reserved

Library of Congress Catalog Card 73-75713

ISBN 8412-0161-7

PRINTED IN THE UNITED STATES OF AMERICA

Advances in Chemistry Series
Robert F. Gould, *Editor*

Advisory Board

Bernard D. Blaustein

Paul N. Craig

Ellis K. Fields

Louis Lykken

Egon Matijević

Thomas J. Murphy

Robert W. Parry

Aaron A. Rosen

Charles N. Satterfield

FOREWORD

ADVANCES IN CHEMISTRY SERIES was founded in 1949 by the American Chemical Society as an outlet for symposia and collections of data in special areas of topical interest that could not be accommodated in the Society's journals. It provides a medium for symposia that would otherwise be fragmented, their papers distributed among several journals or not published at all. Papers are referred critically according to ACS editorial standards and receive the careful attention and processing characteristic of ACS publications. Papers published in ADVANCES IN CHEMISTRY SERIES are original contributions not published elsewhere in whole or major part and include reports of research as well as reviews since symposia may embrace both types of presentation.

CONTENTS

Preface ix

STATUS, PROCESS, AND EQUIPMENT

Introduction 1
    George E. F. Brewer

1. Conversion and Electrodeposited Coatings: A Total Concept 7
    James I. Maurer and Robert M. Lacy

2. New Developments in the Theory and Practice of Pretreatment of Metals Prior to Electrophoretic Coating 38
    Lester Steinbrecher

3. Changes in Zinc Phosphated Steel Surfaces during Electrodeposition 47
    C. A. May

4. Power Supplies for Electrodeposition of Coatings 62
    Claybourne Mitchell, Jr.

POLYMERS AND PIGMENTS

Introduction 78
    George E. F. Brewer

5. A Low Molecular Weight Esterified Copolymer of Styrene and Maleic Anhydride for Electrodeposition 80
    John F. Motier and Donald L. Marion

6. Preparation of Electrodeposition Resins: Lactone Formation during Maleinization 88
    LeRoy S. Forney and Thomas J. Sheerin

7. Pigmentation of Electrocoatings 98
    David S. Young and Arthur T. Gronet

8. Studies in Cathodic Electrodeposition 110
    R. A. Wessling, D. S. Gibbs, W. J. Settineri, and E. H. Wagener

THEORY

Introduction 128
    George E. F. Brewer

9. Electrodeposition of Carboxyl Containing Copolymers. Basic Studies, Phase Growth, Counterion Fixation 130
    A. E. Rheineck and A. M. Usmani

10. Electrochemistry of Polymer Deposition 149
    Zlata Kovac-Kalko

11. Variables Affecting the Kinetics of Polymer Electrodeposition ... 166
    W. B. Brown and G. A. Campbell

12. Dynamic Simulation of the Electrodeposition of Polymers 178
    G. A. Campbell and W. B. Brown

13. Thowing Power as Related to Material Properties with Analysis by Digital Computer Simulation 191
    A. E. Gilchrist and D. O. Shuster

BATH MAINTENANCE AND DESIGN OF MERCHANDISE

Introduction 204
    George E. F. Brewer

14. Material Balance Considerations in an Electrocoating Tank 207
    William Van Hoeven, James E. Lohr, and
    William B. Van Der Linde

15. Turnover Studies on Amino Crosslinked Electrocoating Paints ... 216
    Werner J. Blank

16. Influence of Solvents on the Electrodeposition of Paint 227
    Ivan H. Tsou and Walter Stuecken

17. Improved Corrosion Protection through the Electrocoated Edge Spot-Weld Hem Design 232
    George E. F. Brewer and James W. Mitchell

Index 237

PREFACE

In the late 1950's, as a consultant to Ford Motor Co., I called to the attention of my supervisor, G. L. Burnside, the concept of electrophoretic deposition of water-borne paint compositions. Soon the magnitude of this project became apparent, and our thoughts and experimentations were disclosed in confidence to some leading paint manufacturers. For me, this resulted in a most pleasant cooperation with many farsighted leaders in research and development. Indeed, at that time farsightedness was needed to embark on the project of electrodeposition of paint. Fortunately, the investigations were successful: electrodeposition is now a worldwide operation.

About two years ago, I was asked by the officers of the American Chemical Society Division of Organic Coatings and Plastics Chemistry to preside at the Symposium on Electrodeposition of Coatings at the 161st National Meeting of the American Chemical Society, March-April, 1971, in Los Angeles. The papers presented at this meeting, together with short introductory reviews of the total field, are the subject of the present volume.

Sincere thanks go to the authors of these papers for accepting my invitation to participate in this publication. Their work is published in excellent company. Beyond the authors, credit goes to their supervisors and organizations who have so generously furnished guidance and supporting services which made this volume possible. While I cannot attempt to single out all these indirect contributors in all the organizations involved, my own thanks go to G. L. Burnside, H. N. Bogart, and P. H. Ponta of Ford Motor Co., and last, but not least, to L. J. Nowacki and L. H. Princen of the American Chemical Society, Division of Organic Coatings and Plastics Chemistry.

Brighton, Mich. GEORGE E. F. BREWER
August 1972

Status, Process, and Equipment: Introduction

GEORGE E. F. BREWER

Electrophoretic migration and electrodeposition of colloidal particles have been observed as early as 1809 (1) and are described in the literature as "electrophoretic migration" and "electrophoretic deposition" of clay, waxes, asphalt, rubber, and various colloids from their dispersions in water and in organic liquids (2). For many years, it was apparently difficult to maintain bath conditions which result in continuously repeatable formation of electrodeposits; at any rate, none of these electrodeposition processes seems to have attained commercial reality.

Much work has been done in the field of electrodeposition of organic film formers from aqueous baths. At least three books on the subject have been published in the past few years (3, 4, 5). The electrodeposition of the film former is a radically new form of paint application, which, in conjunction with other operations, results in outstandingly improved corrosion protection. In developing the total process, our interest was directed toward the four major steps (*see* Figure 1):

(1) Metal pretreatment
(2) Electrodeposition
(3) Rinse
(4) Bake

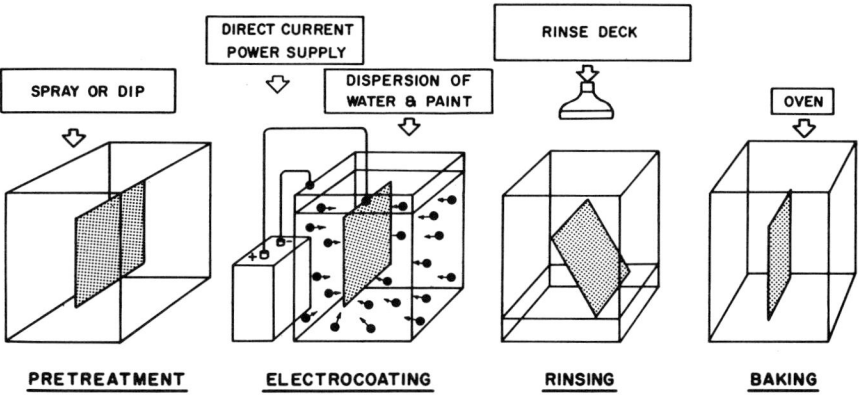

Figure 1. Electrodeposition process

Figure 2. Typical electrocoating tank

To minimize the task of converting existing plants and equipment for electrodeposition, the process was designed to confine equipment changes to one section of the paint department. No change is normally required in the metal preparation equipment and bake oven, which exist in practically all conventional painting operations. All operations subsequent to electrodeposition—such as sanding or top coating—are carried out in a conventional manner.

Electrodepositable paints are on the market for one-coat application as primer-surfacers, etc., in all ranges of gloss and in practically all colors. In these materials, the electrical resistance of the cured coat is usually so high that no second electrocoat can be deposited over the first coat. There are, however, some electrocoating materials on the market from which coats of low electrical resistance are formed, so that a top coat can be electrodeposited (6).

Coating Bath Solids

Most baths contain 5–15% non-volatiles, of which the pigment-to-binder ratio seems to be rarely higher than 1 part pigment to 2 parts binder. A low non-volatile concentration in the bath is desirable since workpieces, when lifted from the bath, carry with them a certain volume of bath. Thus, the lower the bath solids, the smaller the so-called "drag-out loss." Also, lower bath solid concentration results in shorter time of "dwell" in the bath, and pumping stability becomes less critical.

Bath Agitation and Filtration

Paint solid settling is prevented by pumps, draft tubes, line-shaft agitators, ejector nozzles, etc. Their agitational capability is sufficient to cycle the entire tank volume in 6 to 30 minutes. In addition to strainers, filters are provided with pore sizes of 5 to 75 microns. The total bath volume passes through the filter in many cases once every 30 to 120 minutes.

Coating Tanks

Electrocoating tanks (Figure 2), like conventional dip tanks, are operated on either a continuous or a batch type basis. In the latter case, one piece or a number of pieces are located over the tank and then dipped into the bath. A total submersion to 6 inches below the surface with 6 inches clearance on all sides is the usual practice. The tank wall can be used as the counter electrode (Figure 3), or the tank wall can be lined

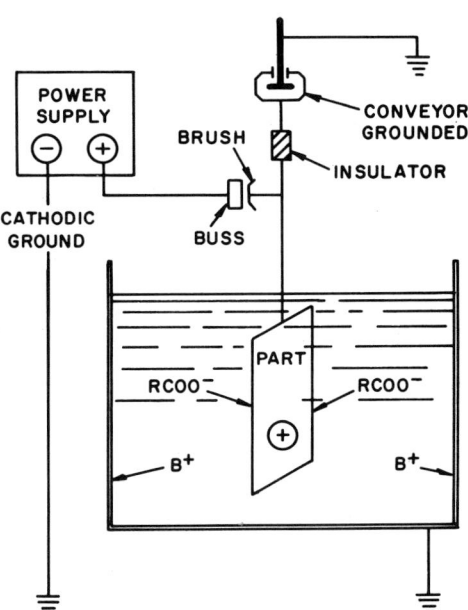

Figure 3. Solubilizer re-use (solubilizer deficient feed)

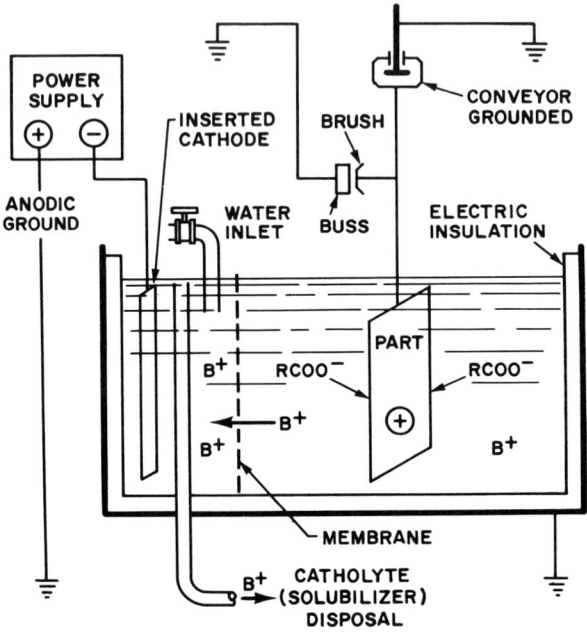

Figure 4. Solubilizer removal (completely solubilized feed)

with an electrically insulating coat (Figure 4), while the counter electrodes are inserted in the tank and then positioned according to size or shape of the workpieces. When lined tanks are used, the workpieces can be grounded through the conveyor. In many cases, however, the hanging device (paint hook) carries an electric contactor (brush) sliding along a grounded rail (buss bar) to ensure electrical ground. If the entire tank wall is used as the electrode, it forms the grounded side of the circuit. In this case, the hook on which the workpieces are hanging carries an insulating link. The lower, insulated section of the hook contacts the buss bar from which it receives the electrical energy.

Cooling Equipment

Practically all the electric input is converted into heat and is removed through adequately sized chillers. Constant bath temperature is usually maintained at 70°–90°F as recommended by the paint supplier.

Electrodeposition

Cleaned or pretreated workpieces enter the electrocoating tank either electrically energized or without electric charge (Figure 5). It is desirable to enter workpieces into a bath at full coating voltage since shorter coating time results. However, some electrocoating baths produce streaks ("hash marks") on merchandise which is entered under full coating voltage. Un-energized or low energized pieces entered into the same bath are free from defects.

Many installations provide the option of un-energized or low energized entry through the use of two or more power sources. Multiple power sources also ensure continued, though lowered, production in case of breakdown of one power source. Amperage limitations, current cycling, or intermittent current application lengthen the required coating time since it is the applied ampere-seconds (coulombs) which produce the electrodeposit.

Current consumption ranges from about 15 coulombs/gram of finished coat up to 150 coulombs/gram, resulting in an overall requirement of 2 to 4 amp/sq ft for 1 to 3 minutes. For special work such as wires, steel bands, etc., coating times as low as 6 seconds are reported. The voltage requirement is largely dictated by the nature of the dispersed resin. Installations are usually operated between 200 and 400 volts though some are reported to operate at voltages as low as 50 volts, others as high as 1000 volts.

Rinse

The electrocoated workpieces, when emerging from the coating tank, carry a well-adhering paint film, almost free from solvent or volatile matter. Bath droplets cling to these surfaces, and difficult-to-drain sections are full of coating bath. The carry out is usually rinsed off or may be reclaimed by a process such as ultrafiltration (7).

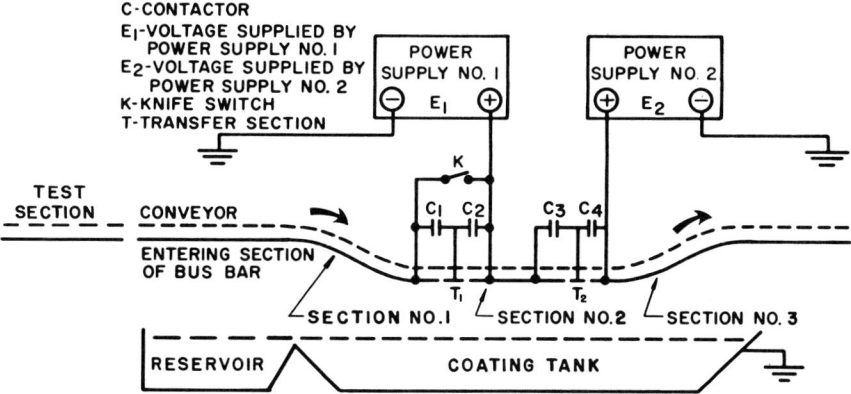

Figure 5. Typical electrical layout for step-up voltage process in electrocoat system

Literature Cited

1. Reuss, *Memorabilia Soc. Imp. Nat. Moscow* (1809) **2**, 327.
2. Hagan, J. W., Ortung, F. W., Chow, S. W., "Electrodeposition of Organic Coating," *Paint Varnish Prod.* (1967) **57** (4) 48–55.
3. Yeates, R. L., "Electropainting," Draper, Teddington, England, 1966.
4. Chandler, R. H., "Advances in Electropainting," London, 1969.
5. Ranney, M. W., "Electrodeposition and Radiation Curing," Noyes Data Corp., Park Ridge, N. J., 1970.
6. Stromberg, S. E., "Conductive Primers and Other EDP Advances," *Ind. Finishing* (Oct. 1969) 22–26.
7. Gerhart, H. L., "Ultrafiltration System Recovers Coatings Lost in Electrocoating Process," *Paint Varnish Prod.* (1971) **61** (2), 40–41.

1

Conversion and Electrodeposited Coatings: A Total Concept

JAMES I. MAURER and ROBERT M. LACY

The Parker Co., Oxy Metal Finishing Corp., Occidental Petroleum Corp., P. O. Box 201, Detroit, Mich. 48220

> *In the electrodeposition of paint, part of the substrate becomes an integral part of the deposited film and can influence the coating properties. This paper covers the effect on paint quality of the cleaning of the metal surface, formation of the conversion coating, post treatment, deionized water rinsing, and dryoff conditions. The system is evaluated for salt spray and humidity resistance, adhesion, filiform corrosion, detergent resistance, and uniformity of paint film. For maximum selectivity of paint, the proper conversion coating, a reactive post treatment, and the dryoff oven should be used. By carefully matching the paint formulation with the conversion coating, quality finishes can result, even if post treatment and the dryoff oven are eliminated. Electrodeposited paints require a more uniform and complete coating than conventionally deposited paints.*

It is accepted practice to clean and to treat metal surfaces to produce on them a conversion coating before applying industrial paint finishes (1, 2). Conversion coatings as a base for paint have been proved valuable during many years of field use, which have shown that they provide a simple, economical means of substantially increasing the overall quality of painted products.

Proper metal preparation, including the formation of a surface conversion coating prior to painting, contributes to painted product durability by:

(1) decreasing the spread of corrosion of the substrate metal at areas where the paint film is broken, and in this way materially reducing the

loss of paint that would ordinarily lift and peel away as a result of the action of the alkaline corrosion products.

(2) preventing or decreasing on zinc surfaces the reaction of the zinc metal with the paint by virtue of the fact that the conversion coating is a non-metallic, non-reactive separating layer.

(3) controlling the action of moisture which permeates the paint film to a substantial degree. This eliminates or minimizes blistering under high humidity conditions, thus contributing to paint film integrity.

(4) improving the mechanical paint adhesion by increasing the surface area, and/or providing a capillary bed (3) into which the organic finish can penetrate.

Conversion coatings are produced by the chemical reaction of a coating solution with the metal surface. In most cases, components of the metal surface react with components of the coating solution to produce a tightly adherent, water-insoluble inorganic coating on the metal. The metal surface is thus rendered non-metallic.

Cleaning and conversion coating can be combined into one step. Processes of this type are generally referred to as cleaner-coaters. Much greater flexibility in operation and usually higher quality can be obtained, however, by separating the cleaning and the conversion-coating step and by post treating the conversion coating to further enhance its ability to hold paint and minimize corrosive attack of the metal surface.

A typical processing sequence before conventional paint application today would consist of:

(1) Cleaning the metal, 60–90 seconds
(2) Water rinse, 30 seconds
(3) Treatment to obtain a conversion coating, 60 seconds
(4) Water rinse, 30 seconds
(5) Post treatment or final rinse, 30 seconds
(6) Dryoff in oven

The two basic methods of treating metal surfaces are by the immersion process and by the spray process (4, 5). The immersion process is the older and the simpler and consists of dipping the product to be treated in tanks containing the treatment solution. Most high production treatments of preformed metal parts today, however use the chemicals by the spray process. The spray process, in this case, is not the application of the solution using finely dispersed particles, as is the case of the application of paints by spray, but rather by flooding the preformed parts by impinging the solution onto the metal surface through nozzles that have relatively high volume capacity and which are designed to produce a minimum breakup of solution. The solution draining from the parts runs back to a reservoir tank and is constantly recirculated onto the work and continually reused. Conversion coatings can be produced by brushing or wiping a treating solution on the metal surface. Portable, heated spray equipment or steam generating equipment can be successfully used to apply conversion coatings outdoors or where spray or immersion equipment is unavailable. Normally these latter methods are used for limited production or for large or heavy items.

One large use of painted metal is the painted coil-postformed approach to the production of items such as roof decking, building siding and trim, and many other applications where the product can be formed after painting. Since the metal is flat and in coil form, it can be pulled through a stripline where each of the processing stages are separated by squeegee rolls. This process features very short treatment times and very high line speeds (6). Both immersion and spray processes are used.

A New Dimension

Until the introduction of electrocoating (7, 8, 9), the method of application of the paint film had little if any bearing on the quality of the finished system. With the introduction of the electrodeposited paint film, however, it is necessary to think no longer in terms of a paint applied on a substrate but rather a paint film formed on a substrate, with components of the substrate becoming an integral part of the paint film (10, 11, 12, 13, 14, 15). With electrodeposited painting it is the interplay of the total finish system that must be considered to ensure optimum balance of quality and economy. This paper deals with the role played by conversion coatings in the total finishing system.

Current Practice

Various conversion coating processes are used in industry today. The type used depends on the type of metal, the combination of metals processed, and the quality requirements of a given operation. The following summarizes the type of conversion coatings available:

Steel:

(1) Iron phosphate—a mixture of iron phosphate and iron oxide; considered to be amorphous. Coating weights in the range of 15 to 90 mg/sq ft.

(2) Zinc phosphate—grey, crystalline, essentially a mixture of zinc and iron phosphates. Coating weights in the range of 100 to 600 mg/sq ft.

(3) Molybdate/phosphate—a mixture of iron phosphate and molybdate with iron oxides; considered to be amorphous. Coating weights in the range of 15 to 50 mg/sq ft.

Zinc Surfaces: (Hot Dipped Galvanized and Electrogalvanized Steel)

(1) Zinc phosphate—essentially zinc phosphate with traces of nickel; grey, crystalline. Coating weight range of 100 to 350 mg/sq ft.

(2) Molybdate/phosphate—a mixture of zinc molybdate and phosphates; bluish-golden in color; amorphous.

Aluminum:

(1) Zinc phosphate—a mixture of zinc and aluminum phosphates; crystalline, grey in color. Coating weight range of 150 to 500 mg/sq ft.

(2) Molybdate/phosphate—a mixture of aluminum molybdate and phosphate; golden-blue in color.

(3) Chromic oxide—oxides of aluminum with chromic compounds; colorless to golden; considered to be amorphous. Coating weight range of 10 to 50 mg/sq ft.

In terms of total metal surface area treated, cold rolled steel is certainly the most important metal painted by electrodeposition. We therefore consider two of the most widely used conversion coatings for this metal—the zinc phosphate and the iron phosphate conversion coatings.

A New Look at the Old

Zinc phosphate coatings are, to the naked eye, light grey, scratchable coatings, with an obvious, fine crystal structure. Under a light microscope, we can see a definite crystalline structure, but it is difficult to obtain a really good view of the coating because the zinc phosphate crystals are translucent to light, and they appear relatively coarse on a micro level. This translucent property, combined with the shallow depth of focus of a light microscope gave the photographs illustrated in Figure 1.

Figure 1. SAE 1010 cold rolled steel surface coated with a zinc phosphate conversion coating using a spray, nitrite-accelerated process. 200 × using a light microscope.

Figure 2. Same SAE as in Figure 1 but viewed at 200 × with a scanning electron microscope

With the advent of the scanning electron microscope (17), which features, among other things, a greater depth of focus and because the coatings are essentially opaque to the electron beam, conversion coatings can now be seen in greater detail.

Figure 2 shows the same coating as in Figure 1 but viewed using a scanning electron microscope. Higher magnification viewing with the scanning electron microscope reveals further details of a phosphate conversion coating on steel.

Figure 3 shows a typical zinc phosphate coating as would be applied to steel or zinc at a magnification of 2000 ×. Figure 4 shows a calcium modified, zinc phosphate process on steel at 2000 ×. Figure 5 shows an iron phosphate coating, which has heretofore been generally considered amorphous, at a magnification of 10,000 ×.

Conversion coating are non-metallic and are generally described as non-conductive (18, 19). Since the electrodeposited paint film can be applied only to a conductive surface, how can the conventional conversion coatings serve as a base for electrodeposited coatings? The answer is that while conversion coatings are essentially non-conductive, they are discontinuous, and the electrodeposition of paint films begins in the pores of the conversion coating (14, 15, 16).

A typical processing sequence for treating metals before conventional painting consists of six stages. For electrodeposited paints, the following sequence is used:

(1) Cleaning the metal
(2) Water rinse
(3) Treatment to obtain a conversion coating
(4) Water rinse
(5) Post treatment or final rinse
(6) Deionized water rinse, 10–15 seconds
(7) Dryoff in oven

Each step is discussed below. A typical spray phosphatizing unit is shown in Figure 6.

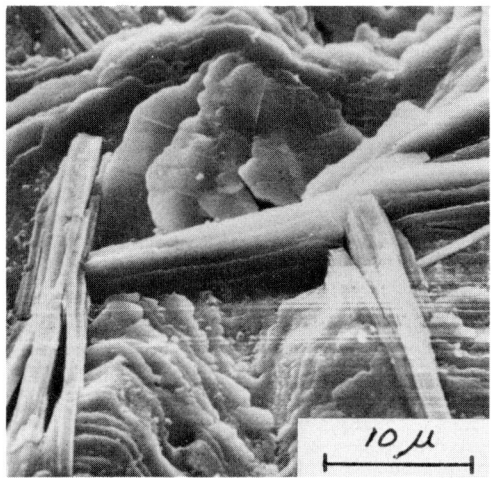

Figure 3. SAE cold rolled steel treated with a spray nitrite-accelerated, zinc phosphate process modified by nickel and fluoride

Cleaners

The first step in forming a conversion coating is attack of the metal surface by the coating solution. To obtain a uniform initial attack, and thus a uniform final conversion coating, all unwanted soils must be removed from the surface of the metal. Almost all metal used in industry is coated with a thin film of oil by the producer to protect it during shipping and storage. To identify various grades of metal, most mills apply a printed identification on their metal sheets or coils, usually giving their tradename and type of metal. Moreover, during annealing and cold rolling, smut-like soils are formed, which consist of mixtures of partially burnt rolling oils, finely divided metal particles, and oxides of the metal.

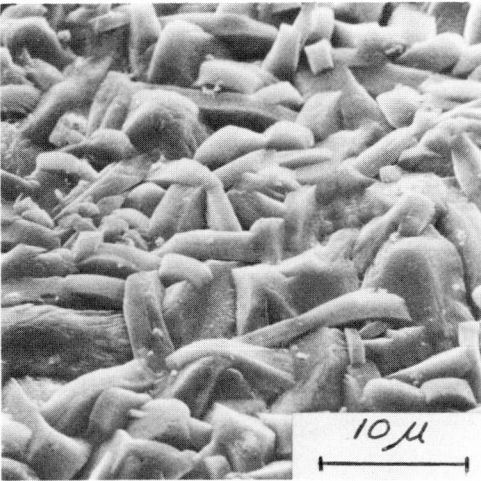

Figure 4. SAE 1010 cold rolled steel treated with a spray zinc phosphate process, calcium modified, ferrous ion present. 200 × using a scanning electron microscope.

To fabricate metal into the desired shape, it is often necessary to apply pressing or drawing lubricants. These are useful because they adhere tenaciously to the metal, reduce friction and wear between the die and the metal, and thus eliminate scoring, scratching, and galling. During

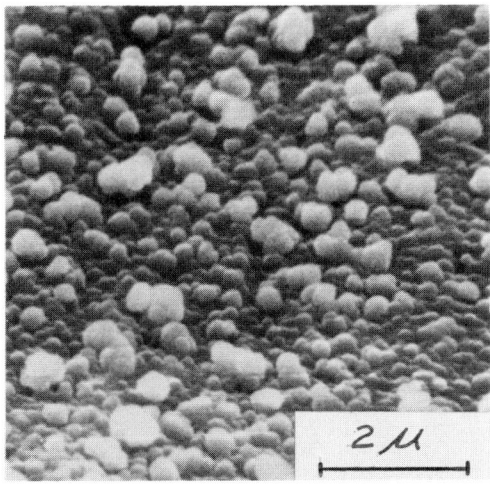

Figure 5. SAE 1010 cold rolled steel treated with a chlorate accelerated, spray, iron phosphate process. 10,000 × using a scanning electron microscope.

storage and during their journeys through fabrication plants the metal parts pick up shop dirt, chalk, wax, and ink which are inspection aids or identification markings. These soils on the metal surface can seriously affect the subsequent metal processing steps and must be removed.

The cleaners formulated and selected must be able to remove the various soils described above, and after the water rinse stage they must produce a surface that is conducive to the conversion coating process.

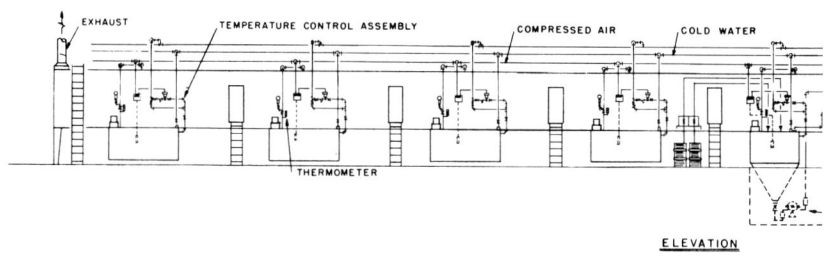

ELEVATION

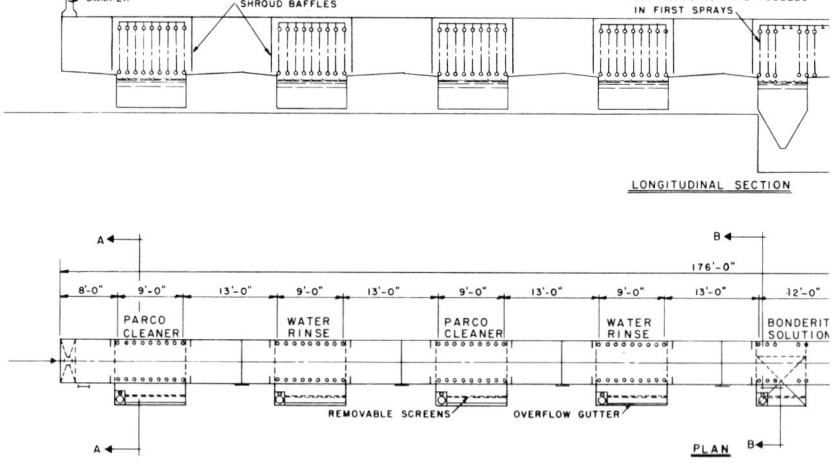

LONGITUDINAL SECTION

PLAN

ZONE	TIME CYCLE	OPERATING TEMP. °F	SCHEDULE NOZZLES		
			NUMBER	TYPE	PRESS
PARCO® CLEANER	45 SEC	150–180	135	H 3/8 U15070	15 PSI
WATER RINSE	45 SEC	150–180	135	H 3/8 U15070	15 PSI
PARCO CLEANER	45 SEC	150–180	135	H 3/8 U15070	15 PSI
WATER RINSE	45 SEC	130–180	135	H 3/8 U15070	15 PSI
BONDERITE® TREATMENT	60 SEC	130–180	105	3/8 BSS 50-50.1	10 PSI
			*12	3/8 KSS 30	10 PSI
WATER RINSE	30 SEC	NO HEAT	105	H 3/8 U15070	10 PSI
PARCOLENE® TREATMENT	30 SEC	NO HEAT	84	3/8 KSS 30	10 PSI
DEIONIZED WATER RINSE	15 SEC	NO HEAT	48	3/8 KSS 30	10 PSI
FRESH DEIONIZED RINSE			24	1/8 KSS 10	10 PSI

*FIRST SET OF SPRAYS IN THE BONDERITE ZONE

Figure 6. Details of

This means that the cleaners must also condition or treat the surface of the metal to render it receptive to the conversion coating step. We therefore speak in industry of cleaning and conditioning the surfaces. Cleaning has always been important in preparing metal for painting. With the introduction of electrodeposited paints, however, the necessity of obtaining uniform conversion coatings has become even more important. With conventional paints, a reasonable degree of uniformity must be main-

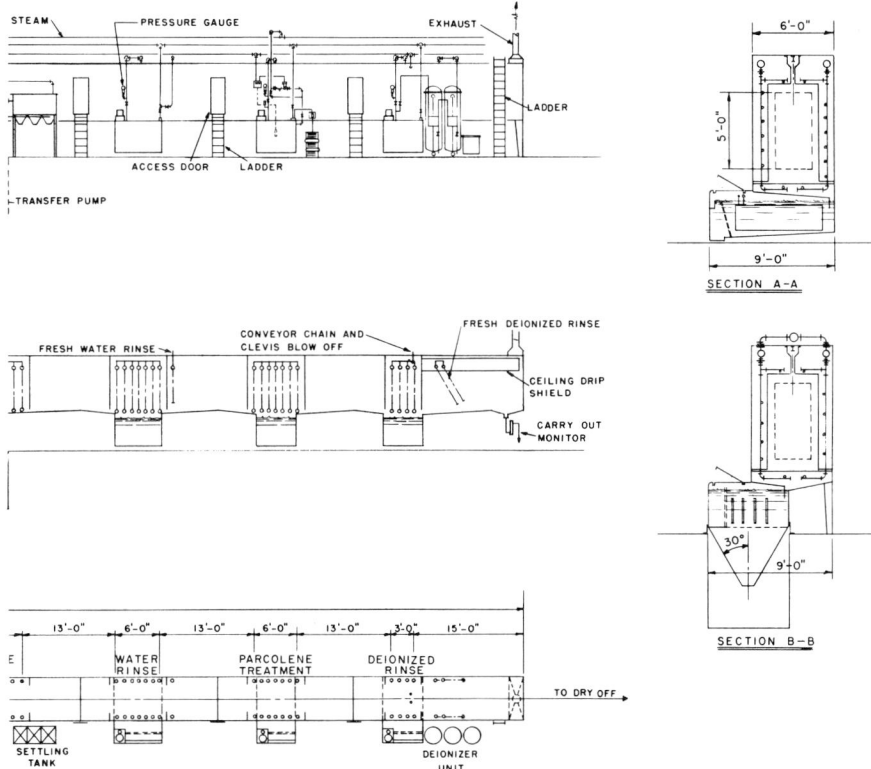

GPM/EA	PUMPS		SOL. TANK CAP. GALS.	MOTORS		BTU/HR NET
	HEAD PRESS	GPM		HP	RPM	
4.3	45 FT	600	1800	10	1750	2,100,000
4.3	45 FT	600	1800	10	1750	1,800,000
4.3	45 FT	600	1800	10	1750	1,800,000
4.3	45 FT	600	1800	10	1750	1,800,000
5.0	35 FT	600	1800	10	1750	1,800,000
3.0						
3.5	35 FT	375	1000	7 1/2	1750	
3.0	35 FT	250	750	3	1750	
3.0	35 FT	225	750	3	1750	
1.0						

spray phosphatizing unit

tained to prevent differential gloss of the paint film. This problem is apparent when a one-coat paint system is applied over a metal part having widely varying crystalline forms of the conversion coating. This problem has not, however, been very serious in the industry, and coatings having visual differences in regard to streak-like discolorations or patterns of different colored conversion coatings are satisfactorily painted without any sacrifice in appearance or quality. The differences, visually apparent, in many conversion coating lines when the cleaning is marginal or weak reflects different size crystals and with electrodeposited paints have been found to reflect differences in apparent conductivity of the surface. Generally speaking, with a zinc phosphate coating, the finer the crystal structure, the less the porosity in the coating. Conversely, the larger the crystal structure, the higher the porosity. The difference in the surface conductivity caused by differences in porosity means that a paint film is applied by electrodeposition at a different rate over one type of coating structure than another; this can result in visual differences that can be undesirable, particularly with one-coat paint systems. It is thus increasingly important that the cleaner formulations be carefully considered with respect to the soil removal and their ability to condition the surface to obtain a uniform conversion coating.

With calcium modified, zinc phosphate processes, a uniform, hard, dense, useful coating can be obtained after strong alkali cleaners. With many other processes, particularly those useful for treating mixed production of zinc and steel or zinc, steel, and aluminum, it is necessary to use a surface activating compound based on titanium phosphate (20). Since the titanium phosphate activator is not stable in highly alkaline solutions, it is necessary in some operations to clean in two stages, separated by a water rinse. In the first stage specially formulated, strong cleaners are used to remove the bulk of the soil from the surface. The relatively clean surface is treated in a milder cleaner containing the activating agent, which in turn produces a surface highly capable of accepting a uniform conversion coating. Another way of handling this situation when the equipment is limited to a single cleaning stage is to use a relatively strong cleaner in the first stage and then inject continuously into the water rinse, a slurry of the titanium phosphate activating compound.

Since industry has become critical about the uniformity of the conversion coating, the synthetic surfactant systems used in the alkali cleaner have a strong bearing on the receptivity of the surface to the conversion coating solution, and a number of cleaners have been formulated that have resulted in improved uniformity in the conversion coating.

In electrocoating, there is a strong tendency for the initial paint deposited to migrate to the outermost skin of the finished surface. With this migration, inerts, smut-like soils, chalk markings, mill identification inks, etc., tend to show up on the finished surface. In the case of primers, this lifting effect may not be significant. However, with one-coat systems, particularly light colors, this can result in irregular color patterns or in readable markings. Normal cleaning techniques may not adequately remove all these smut-like soils, and it is sometimes necessary to resort to mechanical effort (21) such as hand wiping, mechanical brushing, or special solvent touch ups. For hot rolled steel, zinc, aluminum, or steel that has corroded, conventional alkaline cleaners rarely do a satisfactory job. In the case of steel, it may be necessary to remove the scale and rust by conventional acid pickling. With aluminum, deoxidizing steps such as nitric acid or chromate–fluoride treatment may be necessary to remove the oxides.

The best solution to corrosion is its prevention. Therefore, improved in-plant storage is strongly recommended. With conventional paint practices, it is possible to cover up a variety of defects; this is not the case with electrocoating.

Almost all of the cleaners used prior to conversion coatings are alkaline. The formulations vary widely, and most are proprietary. The formulation below is typical of a heavy duty alkali cleaner:

10 to 20% sodium carbonate
15 to 25% tetrasodium pyrophosphate
50 to 70% sodium metasilicate
2 to 8% non-ionic, low foaming, surfactant system

A cleaner such as this would be used at concentrations from 0.5 to 2 ounces per gallon at 66°–77° C (150°–170° F) for 1 to 1½ minutes by spray application. When processing aluminum, it is sometimes desirable to use cleaners formulated primarily of sodium hydroxide to etch the surface of the metal. For aluminum extrusions or stampings that have been scratched or abraded by the forming operation, the etching type aluminum cleaners will tend to smooth out and generally improve the appearance of the part.

Whatever chemical formulation is used, uniform, complete cleaning is necessary since the nature of the cleaning has a strong bearing on the conversion coating step.

Water Rinses

Generally, the chemicals in the cleaner, conversion coating, and post treatment stages of a phosphatizing unit are not compatible. A water-rinse stage is therefore placed between each stage to remove the unreacted chemicals from the metal and thus minimize drag-in of chemicals from one stage to another. Since this is primarily a dilution process, such concepts as the counterflow of the rinse solutions, breaking up the rinse stage into two parts, and introducing the fresh water to the stage through nozzles at the exit of the water rinse stage are used to improve efficiency.

Conversion Coatings

The pioneers in the development of electrodeposited paints had hoped that this revolutionary means of applying a paint film would require only superficially degreased surfaces and would not require conversion coatings for high quality metal finishing (22). As more knowledge and experience were obtained, however, it became clear that proper metal preparation was not only important, but in many ways even more critical than with conventional methods of applying paint. Not only were outdoor exposure resistance, detergent resistance, filiform corrosion resistance, salt fog resistance, humidity resistance, and physical tests influenced by the metal preparation but also the appearance of the paint film in both uniformity and in color. Still another factor that is influenced by the metal treatment is wet film adhesion, which is the ability of the electrodeposited film to resist the water rinsing that it is usually given after the deposition of the paint film, prior to curing.

All of the commercially used electrocoating paint systems apply the paint at the anode. When steel is painted by the anodic electrocoat process, H^+, O_2, and Fe^{2+} are formed at the anode by the electrolytic process. The conventional conversion coatings are essentially nonconductive, and this means that the initial deposition of the paint occurs in the areas between the crystal structure (14). This implies an interrelationship between the structure of the conversion coating, the paint, and the application parameters of a given formulation. That is, depending upon the conductivity characteristics of a given paint system, we will have either higher or lower current densities, which will in turn make for greater or lesser quantities of hydrogen, oxygen, and ferrous iron. In addition to the overall quantities of these materials produced in the deposition of a film, the rate at which they are produced can also have a bearing on the overall results. In turn, the nature of the conversion coating influences both the rate and way the deposit occurs. The migration of the metal dissolved at the anode also is influenced by the nature of the conversion coating.

The interplay between the anodic dissolution with the concommitant formation of gases and metal ions and the characteristics of the deposited paint film also result in a portion of the conversion coating being eroded away and distributed in the resulting film. This means that the paint film is derived not only from the resin and pigment of the paint formulation but also from the dissolved substrate and the conversion coating components. The extent of the conversion coating loss depends on the nature of the coating itself and on a particular paint formulation. The upper part of Table I illustrates the magnitude of this loss. The lower part of Table I shows the salt fog corrosion resistance of the finishing systems (*see also* Figure 7). Obviously there are great differences in the amount of conversion coating loss between various paint systems, and this loss is a function of the particular conversion coating. Also evident is the lack of correlation between coating loss and corrosion resistance. This specific interdependence between the conversion coating and a specific electropaint formulation is one of the interesting characteristics of the electrodeposited painting process. According to May and Smith (*10*), the conversion coating removed during electro-

Table I. Coating Loss and Corrosion Resistance

Conversion Coating	Initial Average Coating Weight Prior to Painting mg/sq ft	Percent Conversion Coating Loss Electropaint Numbers				
		1	2	3	4	5
A[a]	308	10	5	5	6	11
B	198	43	0.6	5	30	16
C	256	44	4	7	11	6
D	65	14	0	0	3	1.5
E	155	75	10	11	21	18
F	129	79	1	12	12	18

5% Salt Fog Corrosion Resistance[b]

		1	2	3	4	5
A	paint creepage	0	1.5–2.5	2–3.5	1–1	0–1.5
	face rusting	2	2	2	3	7
B	paint creepage	0	1–1	1.5–2	1–1	1–1
	face rusting	1	2	1	3	6
C	paint creepage	0	1–1	0–1	1–1	1–1
	face rusting	2	1	1	3	4
D	paint creepage	0–1	78% P	87% P	1.5–3.5	95% P
	face rusting	1	7	3	2	7
E	paint creepage	0.5–1	1–2 7s	1.5–2	1–1	1–1
	face rusting	1	3	3	3	7
F	paint creepage	0–1.5	1.5–2.5	1.52.5	1–1	1–1
	face rusting	2	6	2	3	7

[a] Legend: A—zinc phosphate process, nitrite accelerated.
B—zinc phosphate process, calcium modified.
C—zinc phosphate process, nickel and fluoride modified, nitrite accelerated.
D—iron phosphate process, chlorate accelerated.
E—zinc phosphate process, chlorate accelerated.
F—zinc phosphate process, coating weights controlled by using a dibasic, dihydroxy acid, nickel and fluoride modified, nitrite accelerated.

[b] 336 hours exposure to 5% salt spray (ASTM B117–64), average two panels, paint creepage is the loss of paint from a scribe in 1/16 inch increments or percent. Face rusting is a spot type corrosion of the painted surface; a rating of 1 is perfect; 8 is 100% rusing. See Figure 7 for examples of salt spray failure.

deposition is uniformly distributed throughout the cross-section of the paint film. Cheever and Wojtkoowiak (15) also found that the zinc from the zinc phosphate coating was distributed uniformly on a macro scale throughout the primer that they studied. In contrast to the above, work by Simpson (3), using an electron probe microanalysis technique, concluded that there was no major diffusion or mixing of the conversion coating in the particular primer film he studied but that the removed zinc phosphate coating remained near the primer–metal interface. Simpson produced a zinc phosphate coating on SAE 1010 cold rolled steel, using the conversion coating process C (Table I), applying a primer by electrodeposition and curing at 197° C (395° F) for 25 minutes.

Small strips of the sample were then mounted in a solution of polymethacrylate, which, after hardening, was polished to expose a cross-section of the paint film/steel substrate. The final polish was a 1-micron alumina in water. Following the polishing, the specimens were carefully washed, prior to vacuum deposition of carbon, which was used to dissi-

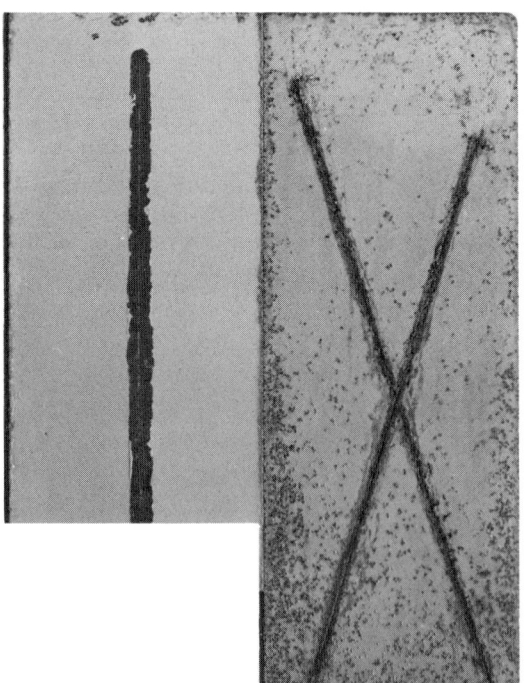

Figure 7. Example of salt spray test results (see Table I)

Left: salt fog creepage for 4-inch wide test panel; creepage 2-4. Right: "face rusting" with "5" rating.

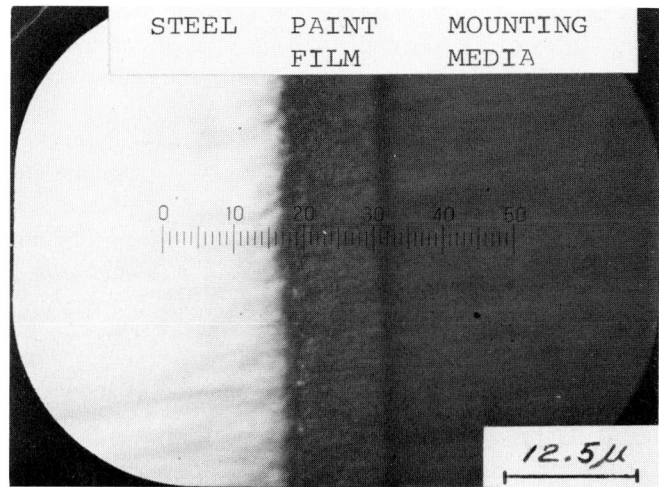

Figure 8. Optical photomicrograph of section of painted steel

pate electric charges during the probe operation. Figure 8 shows an optical photomicrograph of a typical cross section at 800 × magnification.

In the electron probe microanalysis, a 1-micron diameter beam of electrons was scanned across the surface; the resultant and characteristic x-rays were monitored by standard x-ray fluorescence techniques. The specimens were thus scanned across the interface from steel to the mounting media, with measurements of the characteristic x-ray intensities for iron, phosphorus, and zinc. Two typical intensities–distance plots are shown in Figures 9 and 10.

From the data in these figures, for conversion coating C, with this particular primer, Simpson concluded that there was no major diffusion or mixing of the zinc phosphate coating caused by either the anodic dissolution or the turbulent mixing that might have occurred during the heat curing of the paint. Our work in this area, although limited, thus disagrees with the conclusions of May/Smith and Cheever/Wojtkowiak. Possibly some electrodeposited formulations result in more uniform diffusion of the removed conversion coatings than others. Further work needs to be done.

The quality of industrially painted metal is evaluated in different ways. Of course, the ideal way to evaluate a finishing system is to place the finished product in its normal exposure environment and examine it over a lapse of time. Unfortunately, while this approach provides a long range evaluation, it is not a practical means of quality control or a means of developing new products for industrial use. Hence,

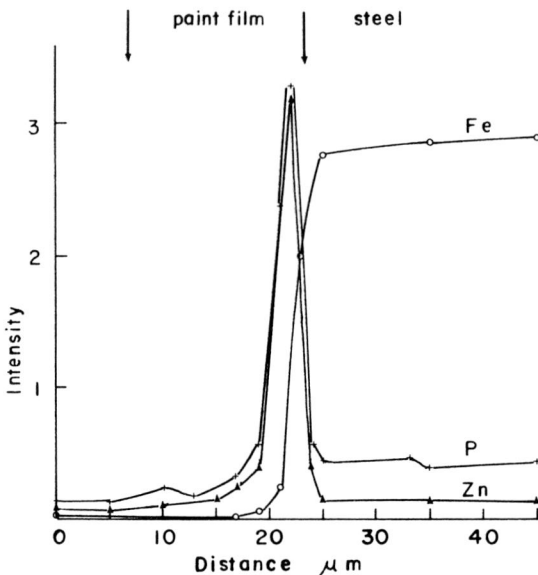

Figure 9. Electron probe scans across painted section. For Fe, scale × 1000 cpm; Zn and P scale × 100 cpm.

laboratory tests have been devised to aid in more rapid evaluation for research and development and quality control. Among the tests which are influenced by conversion coatings are the 5% salt fog test (ASTM B-117), alkaline detergent reistance test (ASTM D-2248), humidity resistance test (ASTM D-2247, 100% relative humidity), and the filiform corrosion (24) resistance test. In addition, various physical tests are used to determine the adhesion of the organic coating system to the metal, such as knife adhesion, deformation tests, the conical mandrel, and impact tests. We must also consider by visual inspection the uniformity and any color changes that might have resulted from the conversion coating on the electrodeposited film.

To illustrate how conversion coatings can and do influence a number of these quality aspects of a finished part, we present selected data covering electrodeposited films on cold rolled steel, galvanized steel, and aluminum. Figure 11 (top) illustrates the degree of difference that can be obtained in salt fog exposure between a cleaned but uncoated cold rolled steel surface *vs.* a steel surface treated to form a zinc phosphate coating. The untreated steel surface has almost completely lost the electrodeposited primer whereas the zinc phosphate treated cold rolled steel surface has retained almost all of the primer. The lower row of panels (Figure 11) shows salt spray corrosion resistance of cold rolled steel surfaces treated with either an iron phosphate (right panel) or a zinc phosphate (left panel) process. Many people have taught that an iron phosphate coating is inferior to a zinc phosphate coating, but with specific electropaints and the proper iron phosphate process results can be obtained that are equivalent to the zinc phosphate process. There are many production lines in operation today using electrodeposited paints that prepare cold rolled steel surfaces by using an iron phosphate process.

Figure 12 illustrates the corrosion resistance of a white electrodeposited paint on aluminum. The test specimen on the left was cleaned and etched whereas that on the right was cleaned, etched, and given a chromic oxide conversion coating. After 7,000 hours in 5% salt spray there is essentially no difference in test results. There are a few pin-point like blisters on the untreated aluminum, but for all practical purposes the cleaned only aluminum surface behaved as well as did the one with the conversion coating. This illustrates that with many alloys of aluminum there is no need for a conversion coating when using electrodeposited paints. One explanation is that there is an *in situ* anodizing of the alumi-

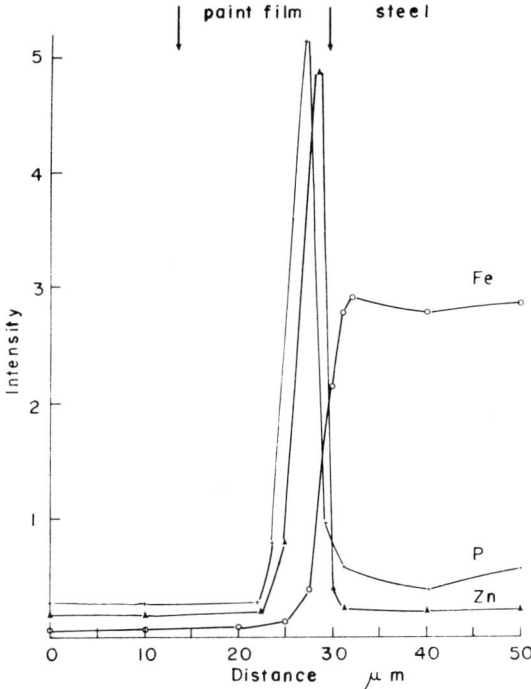

Figure 10. Electron probe scans across painted section. For Fe, scale × 1000 cpm; Zn and P scale × 100 cpm.

num during the initial phases of electrodeposition. Whatever the mechanism, however, in most cases a conversion coating is not needed on aluminum when using electrodeposited paint finishes. A possible exception would be in lines treating a mixture of aluminum with other metal surfaces. If the other metal surfaces, such as cold rolled steel or hot dipped galvanized steel, are treated to produce a conversion coating on their surfaces, the conductivity will be significantly different from that of cleaned and etched aluminum. It may be necessary, under these circumstances to use a conversion coating process that will coat all three metals so that the amount of electrodeposited film will be the same on all three metals. If the conductivity is not the same, the paint film or its gloss will be different.

One insidious type of corrosion that can occur on the surface of painted cold rolled steel, in relative humidities from 50 to 95%, is a thread-like corrosion which has been named filiform cororsion. Conversion coatings on steel surfaces will not stop filiform corrosion, but as illustrated in Figure 13 will substantially decrease it. The panel on the left was cleaned and painted; the panel on the right was cleaned, zinc

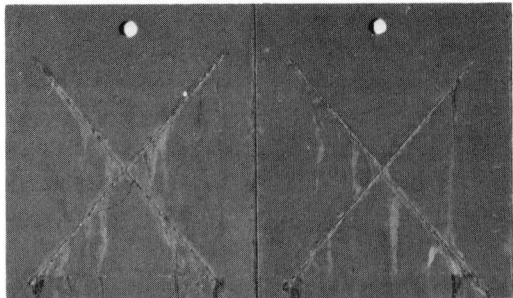

Figure 11. Salt spray corrosion test. Top row: zinc phosphate coating vs. untreated steel; bottom row: zinc phosphate coating vs. iron phosphate coating on steel.

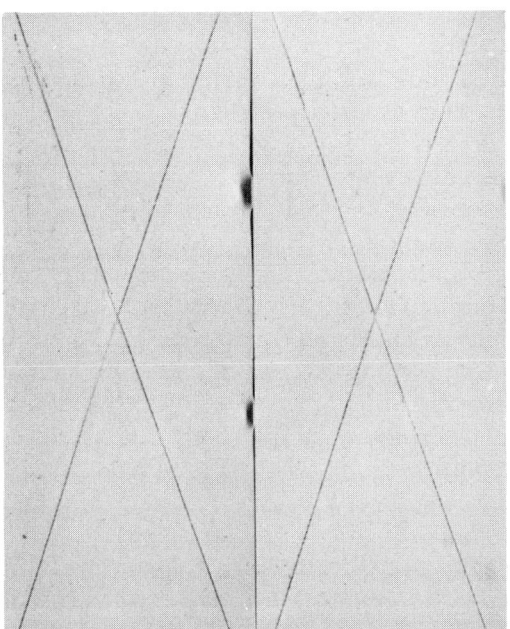

Figure 12. Untreated vs. treated aluminum after salt spray exposure

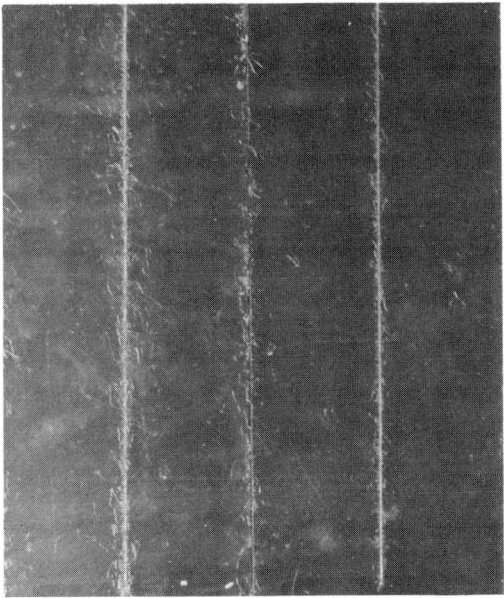

Figure 13. Filiform corrosion. Untreated vs. zinc phosphate coating on steel.

phosphate treated, and painted. The test specimens shown in Figure 14 were exposed for 30 days in a relative humidity of 87% after first activating the test area by exposing the panels for 4 hours in a 5% salt fog chamber.

One of the best conversion coating for steel, where detergent resistance is of prime concern, is the calcium modified, zinc phosphate process. Normally a nickel–fluoride modified, nitrite accelerated zinc phosphate process is not as good as a calcium modified, zinc phosphate process when compared in the detergent resistance test. With the electrodeposited paint formulations designed for detergent resistance, we have found the reverse to be true—the nickel–fluoride, nitrite accelerated zinc phosphate processes give better detergent resistance than do the calcium modified, zinc phosphate coatings. This is illustrated in Figure 14.

Figures 15 and 16 show that it is necessary to select a particular process within a given class. Figure 15 shows two different zinc phosphate processes for hot dipped galvanized steel surfaces; both give excellent test results with conventional paints but show a substantial difference in salt spray corrosion resistance with the particular electrodeposited

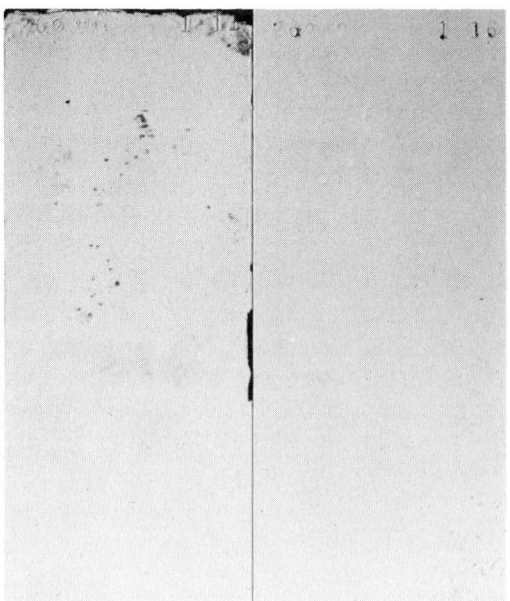

Figure 14. Detergent test results. Calcium modified zinc phosphate (shown on left) vs. nickel-fluoride modified zinc phosphate coatings (shown on right).

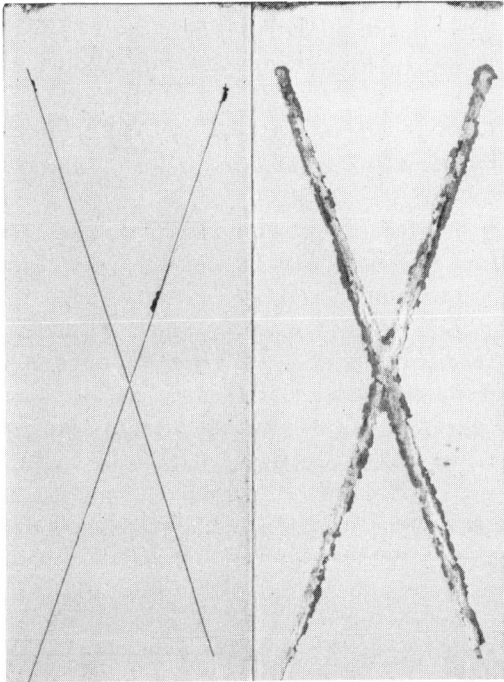

Figure 15. Salt spray resistance. Two types of zinc phosphate coatings on hot dip galvanized steel.

paint formulation used. The same type of difference is illustrated in Figure 16 which compares two different zinc phosphate treatments on cold rolled steel. Again we see a difference with the electrodeposited paint formulation, whereas with a conventional paint we would expect to see little if any difference between these two processes.

These illustrations show that it is impossible to make general recommendations as to the proper conversion coating process to use for a metal surface without a study of the particular electrodeposited paint formulation that will be used.

Post Treatment, Deionized Water Rinse, and Dryoffs

Conventional Paints. Normally, conversion coatings for conventional paints are post treated with a dilute chromate rinse and then dried. The chromate post treatment can be as simple as a 0.1% chromic acid solution or as complex as combinations of calcium chromate, calcium phosphate, or chromic chromate rinses with controlled pH's. The chromate rinses reduce the tendency of paints blister under exposure to high humidity conditions and help to improve the corrosion resistance as resistance as measured by the salt fog test and outdoor exposure. Unfortunately, however, in areas where the chromate rinses accumulate (such as the bottom edges of parts, crevices, and around holes) a point is reached where the blistering occurs, either in spite of or because of too high a concentration of chromate salts.

To eliminate the problem of salt accumulation, the production items are rinsed with an essentially ion-free water (deionized) water. The deionized water rinse washes off the chromate and water salts, thus increasing the humidity resistance of the finish. However, until recently, the corrosion resistance was decreased by this rinse. The problem of decreased corrosion resistance resulting from the use of a final deionized water rinse was recently solved by the introduction of the chromic–chromate post treatment (U.S. patents 3,222,226 and 3,279,958). The chromic–chromate post treatment, operated as taught in the reference patents, allows the removal of the soluble salts from the surface by the deionized water rinse without diminution of the corrosion resistance. With conventional paints, little if any difference in quality is obtained as a result of the type of drying in the range of room temperature up to an oven temperature of 260° C, and from 5 to 10 minutes. The essential requirement with conventional solvent-based systems is that the part be free of surface water before the finish is applied.

Electrodeposited Coatings. Electrodeposited finishes use a water-based system that is sensitive to electrolyte content. Since the electropaints are water based, there should be no reason why the ware has to be dry prior to entry into the paint. The obvious advantages of going in wet, with the elimination of costly dryoff ovens, the cost of fuel to operate the dryoff ovens, and the elimination of the floor space for both the oven itself and the cooling area necessary to reduce the temperature of the part prior to entry into the paint tank have prompted extensive study into the interrelationship among post treatment, dryoff, and the paint formulations.

Since electrodeposited paint systems are sensitive to electrolyte content, it is almost always necessary to rinse the conversion coatings with deionized water to decrease the electrolyte drag in to the paint tank to a point where it does not cause any long term problems. At first, it was suggested that the chromate salts from the post treatments were the biggest problem to the paint. Since many primer formulations contain chromates, however, it is apparent that chromates *per se* are not the principal cause of paint problems, except for their contribution to the total electrolyte content of the paint.

The exact amount of electrolyte that can be safely carried into the electrodeposition paint tank will depend upon the rate of turnover of the paint, the relative drag-out of the paint, which is a function of the part shape and drain time, and the nature of the paint formulation. Close cooperation is necessary between the paint supplier and the operator of the plant to ensure adequate deionized water rinsing in commercial practice.

We believe the most efficient approach is to provide a thorough rinse by a recirculating deionized water zone, followed by a fresh deionized water spray. Figure 6 shows a final deionized water stage as recom-

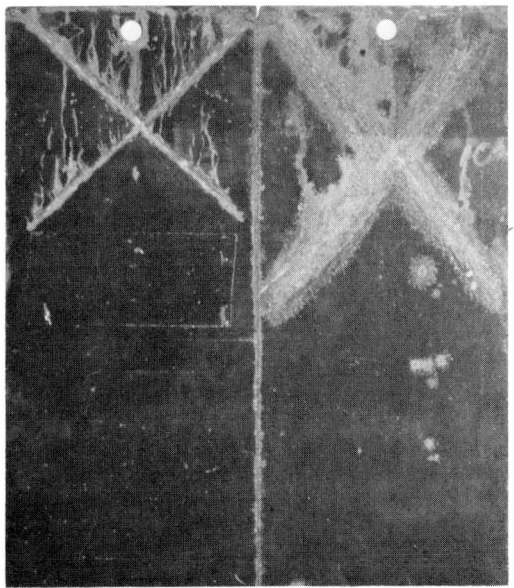

Figure 16. Salt spray resistance. Two types of zinc phosphate coating on steel.

mended by The Parker Co., Oxy Metal Finishing Corp. Since the effectiveness of the rinsing step depends upon the quantity and quality of the water as well as the effectiveness of spraying it to reach all parts of the work, it is advisable to determine the electrolyte content of the water actually carried into the paint. The collection of the rinse drippings from the work and the continuous measurement of their conductivity by a carryout monitor (*see* Figure 6) provide a simple, effective means of ensuring that a predetermined maximum level of electrolyte input will not be exceeded.

While the conductivity range of rinse drippings will vary with the paint turnover and type of paint, a useful point of departure has been to adjust the deionized water rinsing such that the drippings have a maximum conductivity (13) of 60 micromhos cm^{-1}. If we accept this value and the need for the deionized water rinse to ensure continued high performance from the paint, we might ask if any value is derived from a post treatment with a chromate-containing solution when applying electrodeposited finishes. With post treatments, other than those covered by U.S. patents 3,222,226 and 3,279,958 (that is, rinses not operated with trivalent chromium and with pH's outside the range of 3.8 to 6), we find little if any quality improvement results from their use.

The need of or benefit derived from the post treatments covered by the subject patents are a function of the paint systems, the presence of or the lack of a dryoff step, and the type of conversion coating over which they are used. The chromic–chromate rinses used as taught under the subject patents will hereafter be termed reactive post-treatments.

The degree of drying has little, if any, effect on conventional solvent-based paint systems. With electrodeposited paints the nature of the dry-

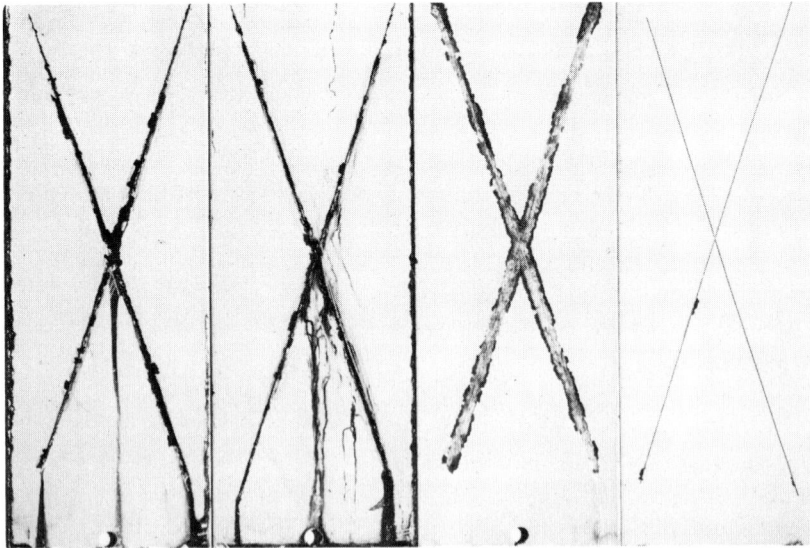

Figure 17. Effect of chromate post treatment on salt spray results when painted wet

Test conditions reading from left to right: (1) steel, zinc phosphate, deionized water, painted wet; (2) steel, zinc phosphate, reactive chromate, deionized water, painted wet; (3) hot dipped galvanized steel, zinc phosphate, deionized water, painted wet; (4) hot dipped galvanized steel, zinc phosphate, reactive chromate, deionized water, painted wet

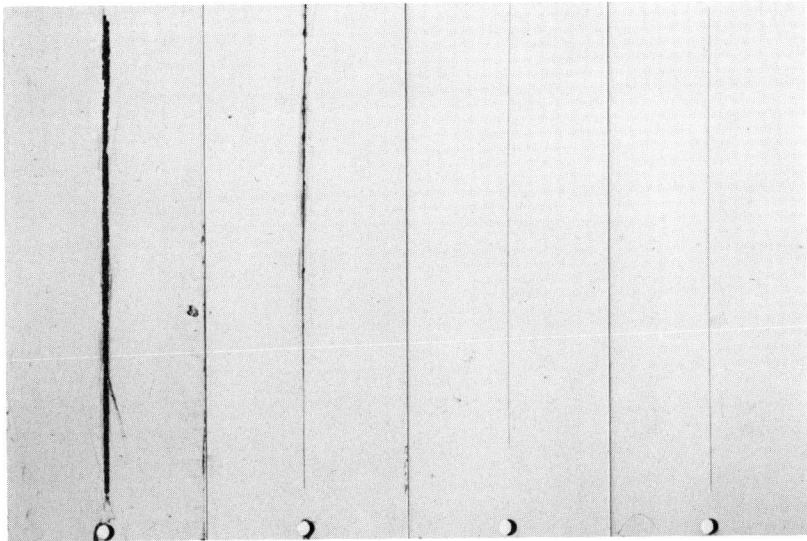

Figure 18. Effect of chromate post treatment on salt spray results when painted after oven drying

Test conditions reading from left to right: (1) steel, zinc phosphate, deionized water, oven dried; (2) steel, zinc phosphate, reactive chromate, deionized water, oven dried; (3) hot dipped galvanized steel, zinc phosphate, deionized water, oven dried; (4) hot dipped galvanized steel, zinc phosphate, reactive chromate, deionized water, oven dried

off can have a significant effect on the end quality. The interplay of post treatment, dryoff oven, and paint is most apparent with salt spray corrosion resistance, wet film adhesion, and whiteness of white, one-coat paints.

In general, all electrodeposited paints are improved by reactive post treatment and a dryoff oven. The degree of improvement varies tremendously, however, with the paint formulation and the type of conversion coating.

Figures 17, 18, 19, and 20 show the salt spray corrosion resistance of a number of electropaints as a function of the post treatment and dryoff. To illustrate the differences that can be obtained, data have been selected using zinc phosphate on cold rolled and hot dipped galvanized steels.

Figure 17 is an example of the effect on salt fog corrosion resistance using a zinc phosphate coating with and without a reactive post treatment with a specific electropaint formualtion. The two panels on the left are cold rolled steel; the two panels on the right are hot dipped galvanized steel. With the cold rolled steel, there is essentially no difference in corrosion resistance as a function of post treatment, whereas with the galvanized steel, there is a significant difference in corrosion resistance; the test specimen without the reactive post treatment is much weaker than the test specimen treated with the reactive post-treatment. In contrast to the results in Figure 17, Figure 18 shows a similar test with a different electropaint. In this case we see no difference in the corrosion resistance of the hot dipped galvanized steel with or without a reactive post treatment while we do see a difference in the corrosion

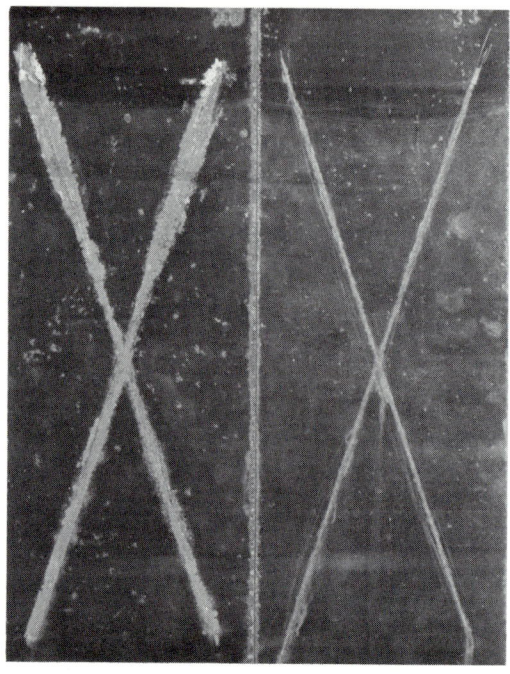

Figure 19. Salt spray test results with given electropaint as function of heating of conversion coating

Left: steel, zinc phosphate, dried at 149°C (300°F)

Right: steel, zinc phosphate, dried at 260°C (500°F)

resistance of the cold rolled steel surfaces. Figure 19 shows the changes that can occur in corrosion resistance when the conversion coating is heated to drive off three of the four waters of hydration. With this particular electropaint, we obtained an improvement even at this relatively high dryoff condition. Figure 20 illustrates that the effect of dryoff varies with the conversion coating. Here, all four panels have been coated with the same electropainted finish; however, the upper row of panels has one type of zinc phosphate coating on it, the lower row a different zinc phosphate coating. We see little difference between the corrosion resistance when the coatings are oven dried prior to the application of the paint film but a considerable difference in corrosion resistance when the paint film is applied on these wet conversion coatings.

The reason for a change in the characteristics of an electrodeposited paint film because of heating of the conversion coating is not fully understood. One possible explanation appears to be found in a study of the dehydration of the zinc phosphate coating as a function of temperature and time. Zinc phosphate coatings have as their principal constituents two minerals, phosphophyllite ($Zn_2Fe(PO_4)_2 \cdot 4H_2O$) and hopeite ($Zn_3(PO_4)_2 \cdot 4H_2O$) (25). An extensive study of the dehydration of zinc

Figure 20. Effect of dryoff. Top: one type of zinc phosphate coating; bottom: another type of zinc phosphate coating.

Upper left: steel, zinc phosphate, reactive chromate, deionized water rinse, painted wet

Upper right: steel, zinc phosphate, reactive chromate, deionized water rinse, oven dried at 149°C (300°F) prior to painting

Lower left: steel, zinc phosphate, reactive chromate, deionized water rinse, painted wet

Lower right: steel, zinc phosphate, reactive chromate, deionized water rinse, oven dried at 149°C (300°F) prior to painting phosphate conversion coatings was carried out in our laboratories in 1961. Figure 21 shows typical data obtained from this study (26). The process of dehydration is at least partially reversible, and the unpainted zinc phosphate conversion coatings will regain lost water from the atmosphere.

Whether zero, one, two, or three molecules of water of crystallization need be removed for maximum corrosion resistance will depend upon a given paint and the particular type of zinc phosphate conversion coating. In general, removal of two molecules of water of crystallization, which would occur in many standard dryoff ovens, is adequate. However, this depends on the paint and the conversion coating; some paints improve by removal of a third molecule of water; others give satisfactory results when painted wet.

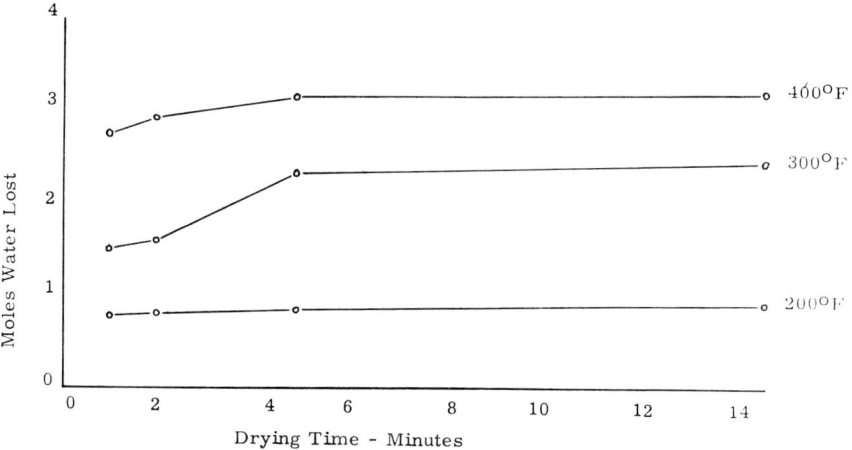

Figure 21. Dryoff vs. water of crystallization loss as function of temperature

In Europe and in Japan, because of increased pressures to minimize pollution, the trend has been to eliminate the post-treatment step in preparing metal prior to electrodeposited paints. This has been done because of their lack of disposal facilities to handle chromate containing wastes or their disinterest in providing them. In both Europe and Japan, energy to produce heat has always been at a premium and therefore they have been among the first to eliminate the dryoff oven step. Under production operations, using neither a reactive post-treatment nor a dryoff oven, they have experienced poor, wet film adhesion. By this we mean a loss of the paint during the water rinsing of the electrodeposited paint film prior to curing. This loss of adhesion can be quite spotty on a given part and apparently occurs generally in the areas of lowest current density. The problem of poor, wet film adhesion is definitely related to the paint formulation and the amount of electrolyte content. Figure 22 illustrates this problem. However, with everything else held constant, the problem becomes more prevalent when neither a reactive post treatment or a dryoff oven is used. In most cases, the use of either the reactive post treatment or a dryoff oven will significantly improve or eliminate the problem.

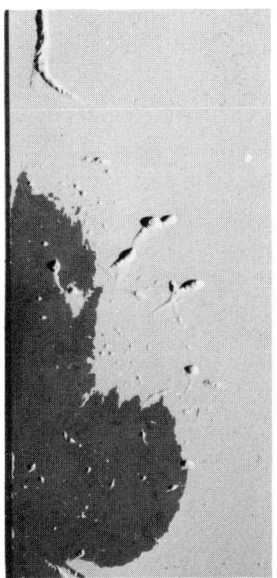

Figure 22. Example of poor wet film adhesion

While pollution pressures have forced some to be more concerned with pollution of the environment than we have been, we are awakening to the problem, and there is no doubt that the most competitive total system using electrodeposition would be one that produces a conversion coating, that when matched with an electrodeposited paint provides acceptable quality with the elimination of the post-treatment stage and the dryoff stage.

This paper is concerned with the effect of eliminating the post-treatment and the dryoff oven as it relates to the conversion coating system. There are other reasons to eliminate the dryoff oven or to avoid the elimination of the dryoff oven, based on considerations such as the presence of the sound deadener pads in the case of the automotive bodies. By eliminating the dryoff oven, the sound deadening pads go into the electrodeposited paints full of water and therefore do not pick up any significant amount of paint, thus reducing overall costs. In some cases, moisture or water in crevices may result in a dilution, in local areas, of the electrodeposited paints, resulting in less than desirable coating deposition.

The electrodeposited one-coat white systems over aluminum or galvanized steel surfaces, if properly cleaned and treated, do not present a serious problem insofar as obtaining a uniform white surface is concerned. However, the electrodeposition of one-coat systems on steel is another matter. Because of the anodic dissolution of the substrate, a degree of yellowing or tan discoloration of the paint film is obtained from the iron salts. Even with the best conversion coatings for steel, there is a tendency to discolor the paint if the reaction post treatment and the dryoff oven are omitted. It appears that the passivity obtained by the reactive chromate treatment goes a long way to minimize either the diffusion of the dissolved iron through the paint film or its ability to react with the paint components. Whatever the mechanism, we have found that the reactive post-treatments help assure the production of a more uniform and white, electrodeposited paint systems.

Conclusions

It is necessary to recognize the interdependence of the options available in metal treatment and electrodeposited paint formulations. For maximum flexibility with respect to the selection of electrodeposited paints, the proper conversion coating, reactive post treatment, and dryoff oven should be used. By matching the electrodeposited paint formulation with the conversion coating, quality finishing can be obtained even with the elimination of the reactive post treatment and the dryoff oven. Tighter control of the electrodeposited paint and painting conditions are required when these two steps are not used.

Present metal treatments result in high quality finishing systems. The metal treatment industry cannot be complacent, however, because of the increased awarness and concern for eliminating the pollution of our environment, it is essential that metal treatment—electrodeposition systems be developed that provide quality finishes without pollution. Development of such systems will require the cooperation and full understanding of all parties concerned—those working in metal treatment, paint formulation, and equipment design and finally the user himself.

Literature Cited

1. "Metals Handbook," Vol. 2, 8th ed., p. 529–547, American Society For Metals.
2. Cavanagh, W., Gibson, R., "Phosphate Coating of Metal Surfaces For Industrial Use," *Plating* (June 1955).

3. Cheever, G. D., "Wetting of Phosphate Interfaces by Polymer Liquids," "Interface Conversion For Polymer Coatings," Weiss, P., Cheever, G. D., Eds., Elsevier, New York.
4. Maurer, J. I., "Preparation of Metal Surfaces for Organic Finishes," American Society of Tool and Manufacturing Engineers, paper FC68-652.
5. Maurer, J. I., "Surface Preparation for Organic Finishes," *Ind. Finishing* (April 1969).
6. Ellis, J. W., Maurer, J. I., "Metal Preparation for The Coil Coating Industry," *J. Paint Technol.* (July 1967) **39**, 460–463.
7. Bogart, H. N., Burnside, G. L., Brewer, G. E. F., "The Concept and Development of The Ford Electrocoating System," Society of Automotive Engineers, paper 988A.
8. Revelt, P. A., "Goodbye To Rust and Corrosion," *Ward's Auto World* (1970) **6** (2).
9. LeBras, L. R., "Electrodeposition—Theory and Mechanisms," *J. Paint Technol.* (Feb. 1966) **38**, (493).
10. May, C. A., Smith, G., "Dissolution of The Anode During The Electrodeposition of Surface Coatings," *J. Paint Technol.* (Nov. 1968) **40** (526).
11. "Conversion Coatings Ready For Electropainting," Steel Magazine (Sept. 1966) 46, 47.
12. Maurer, J. I., Saad, K. I., "Pre-Cleaning Determines Electropaint Finishes," *Canadian Paint Finishing* (July 1967).
13. Saad, K. I., "How to Prepare Metal Surfaces for Electropainting," *Product Finishing* (May 1969) 46–55.
14. Hays, D. R., White, C. S., "Electrodeposition of Paint: Deposition Parameters," *J. Paint Technol.* (Aug. 1969) **41** (535).
15. Cheever, G. D., Wojtkowiak, J. J., "Instrumental Studies of The Surfaces and Internal Composition of Paint Films," *J. Paint Technol.* (July 1970) **42** (546).
16. Menzer, W., "Chemische Oberflächenbehandlung von Metallen vor der Electrotauchlackierung," Internationale Tagung für Oberflächentechnik der Metalle, Hannover, May 5-8, 1968.
17. Kimoto, S., Russ, J. C., "The Characteristics and Application of The Scanning Microscope," *Mater. Res. Standards* (Jan. 1969).
18. Maher, J. F., "Phosphate Coatings," *Metal Finishing*, 1970 Guidebook, pp. 594, 596.
19. Machu, W., "The Kinetics of The Formation of Phosphate Coatings," "Interface Conversion For Polymer Coatings," Weiss, P., Cheever, G. D., Eds., p. 130, Elsevier, New York.
20. U.S. Patents **2,310,239** and **2,874,081**.
21. Adams, M., "Cleaning and Phosphating of Assembled Bodies," Society of Automotive Engineers, Paper 668A.
22. Ellinger, M. L., "Electrophoretic Deposition of Paints in Further Fields of Metal Finishing," *J. Paint Technol.* (March 1969) **41** (530).
23. Simpson, V. P., Hooker Chemical Research Center, Grand Island, N. Y., private communication.
24. Van Loo, M., Laiderman, D. D., Bruhn, R. R., "Filiform Corrosion," (Aug. 1953) **9** (8).
25. Laukonis, J. V., "The Role of Oxide Films in The Zinc Phosphating of Steel Surfaces," "Interface Conversion for Polymer Coatings," Weiss, P., Cheever, G. D., Eds., Elsevier, New York.
26. Maurer, J. I., Saad, K. I., "Study of Coatings Prior to Electrodeposition of Paint," Parker International Conference, Detroit, Mich., Oct. 1966.

RECEIVED May 27, 1971.

2

New Developments in the Theory and Practice of Pretreatment of Metals Prior to Electrophoretic Coating

LESTER STEINBRECHER

Amchem Products, Inc., Ambler, Pa. 19002

> *Any conversion coating system for metal parts to be electrocoated must be adapted to the entire ultimate paint system, and the optimum operating conditions must be determined by experimentation. Zinc phosphate and iron phosphate coatings are the two recommended systems for steel. Both suffer stripping during electrocoating, the specific amount of weight loss varying with the pH of the paint and the applied voltage. Weight losses of about 5–10% of the original coating are not uncommon. Poor "wet paint adhesion," which on rare occasions is evidenced by removal of part of the electrocoat film during water rinsing, has been tied in with excessive coating weight loss. The weight of the zinc phosphate coatings applied can vary up to 500 mg/sq ft. Thin phosphate coatings are less corrosion resistant but allow better paint gloss and provide greater resistance to adhesion loss as in stone chipping. A zinc phosphate coating system has been developed which eliminates the problem of discoloration of white electrodeposited paints.*

Each electrophoretic paint system must be checked under various types of pretreated surfaces to ensure that the paint system is compatible with the coated surface and produces optimum corrosion resistance and physical properties. This novel method of paint application has motivated researchers in the field of prepaint treatments of metal surfaces to investigate existing and entirely novel processes which offer optimum substrates for the many different types of electrophoretic paint systems. Furthermore, solutions to problems which are specific to electropainting has forced us in some cases to require extremely close control of all pretreatment stages and formulation of new products to deal with specific situations. This paper deals with the methods of evaluating pretreatments of metal parts to be electropainted, phosphate stripping during electrodeposition, optimum methods of cleaning, water rinsing, phosphating, passivation rinsing, effects of drying at different temperatures prior to electropainting, and rinsing after painting.

Evaluation Procedure

Before an optimum prepaint treatment recommendation can be offered, we had to test the particular paint in question in our electropaint laboratory. Panels of $4'' \times 12''$ commercial grade steel and galvanized steel are treated, utilizing sequences of cleaning and phosphatizing, subjected to different conditions of final rinse, such as no rinse, DI water (DI = deionized), chromate and non-chromate passivating rinses, and passivating rinses followed by DI water. Furthermore, each of these variations is then air dried, baked, or left wet before entering the electropaint bath. The panels are then electropainted in the laboratory apparatus which consists of the following: a dc power supply which operates at a maximum of 1000 volts and 25 amps and a 15-liter stainless steel storage vessel for paint which is maintained at a constant temperature in a water bath. Sometimes the panel is lowered into the paint bath, and the voltage is applied slowly, never allowing the current to exceed 3 amps. At other times the panel is lowered into the bath at a controlled rate with power on to simulate specific production conditions. The panel is removed from the paint bath, rinsed with DI water, air dried, and oven baked at the suggested temperature and time. The phosphating is performed in the same laboratory so that aging of the coating does not become a variable in the testing and closely simulates line conditions between the time of phosphating and electropainting. The painted panels are then tested for salt spray resistance, humidity, various adhesion properties, and gravelometer resistance.

Phosphate Stripping during Electrodeposition Studies

In general, results indicate the dissolution or peeling of zinc phosphate coatings during the electrodeposition varies with the pH of the paint and the voltage applied. Paints operating in the lower pH range of 7.4 to 7.8 have yielded highest coating weight losses from 25 to 50%. Higher pH systems, above 9.0, have yielded the lowest losses of 0 to 3%. For a given paint system, halving the voltage reduced the amount of coating lost by a factor of 3. Reproducible phosphate coating weight losses will be obtained only if a certain procedure of testing is strictly followed.

After studying many paint systems originating from all over the world, it was found that the zinc phosphate which is dissolved becomes an integral part of the paint coating itself. Chemical analyses of used paint baths in our laboratories have shown the complete absence of zinc phosphate in the paint bath proper.

On rare occasions paint systems have been encountered which show extremely poor adhesion to the zinc phosphated surface when the electrodeposited film is water rinsed immediately after painting. The paint is either completely or partially removed by the spray water rinse or can be easily removed by gentle rubbing. When this phenomenon occurs, the zinc phosphate coating is almost completely removed. The pH of the paint system does not appear to be a significant factor during this severe stripping of the zinc phosphate coating. The theoretical factors causing this problem are at present unknown to us, but we believe that the entrapment of this large amount of zinc phosphate in the paint film itself is a partial cause of this poor wet adhesion. Our laboratories have found procedures to minimize or completely correct this problem (discussed later).

Occasionally one comes across the statement in the literature that iron phosphate coatings are not dissolved during the electropainting operation. If this were true, iron phosphates could be used successfully as substrates to electrodeposit white paints. Actually, iron phosphate-coated steel shows more discoloration in a white electrodeposit than zinc phosphated steel. Since iron phosphates have coating weights on the order of only 20 to 40 mg/sq ft, a 10% weight loss during painting would amount to only 2 to 4 mg/sq ft compared with about 15 to 30 mg/sq ft loss in the case of zinc phosphate coatings. Thus, less phosphate coating is incorporated in the paint film in the case of iron phosphates.

If a particular paint system is operated below the rupture voltage, there does not appear to be any correlation between the amount of coating lost during electrodeposition and the corrosion resistance of the paint.

Cleaning

The use of a proper cleaner is essential for quality. The most important factor to consider is that the cleaner is compatible with the phosphating processes and that the surface be free of soil, oil, and solid dust particles so that it can accept a uniform and tight coating. Any irregularities in the coating appearance (streaks, oil stains, or rust or dust) will be magnified during the electropainting operation.

When aqueous strong alkaline solutions are used for cleaning prior to a zinc phosphate process, it is essential for suitable appearance and high quality performance that a grain refining material containing titanium phosphate be used in the stage preceding the zinc phosphate bath so that a uniform fine crystal will be obtained. Where milder cleaning at a pH below 10 can be used, titanium base may be incorporated in the cleaner proper. Strong alkali cleaning, followed by a separate refining stage is preferred. In the case of iron phosphate processes strong alkaline cleaners are preferred and titanation is unnecessary.

As a result of the strong alkali treatment, pH 12 or higher, numerous nuclei are inactivated because they are covered by oxides or hydroxides while sulfuric or hydrochloric acid pickling may destroy up to 90% of the active centers. By treating the metal surface with a titanium phosphate solution, or with similar activating solutions, numerous new crystal nuclei are created at the local cathode, upon which the phosphates can then grow. This crystal growth proceeds particularly easily upon crystal nuclei, which themselves consist of phosphates. The larger the number of nuclei, the less is the distance of the nuclei from the metal surface. The crystals of the heavy metal phosphate will then be very closely packed so that the numerous crystals will soon contact each other. Therefore, fine-grained and thin phosphate coatings will be obtained. Coarse-grained zinc phosphate coatings which result after strong alkaline cleaning or acid pickling and no titanium activation always have a greater area of pores than fine grained zinc phosphate coatings. A greater number of pores present in a coarse zinc phosphate coating produces a more conductive surface and consequently a higher paint film thickness than that produced on a fine-grained surface. If a steel panel is cleaned in a strong alkaline solution and the bottom half is treated in a dilute titanium phosphate solution, zinc phosphated, and then electropainted, the interface between the coarse and refined area will be easily seen and the paint film thickness will be greater in the coarse area. Thus, a non-uniform zinc phosphate coating will result in a non-uniform electropaint film.

Work which has been previously derusted in areas with a phosphoric acid deruster is left with a very light iron phosphate film in these areas. If this work is then passed through a power spray washer and zinc phosphated, the iron phosphate areas resist the zinc phosphate solution. The resulting phosphated part appears with patches of amorphous iron phosphate surrounded by zinc phosphate crystals. If this part is now electropainted, the areas of iron phosphate will show a much higher film thickness than the surrounding zinc phosphated surface and might actually show signs of rupture in these areas. Therefore, non-phosphoric acid derusters should be used if the part is to be subsequently zinc phosphated and electropainted.

Phosphating

The choice between zinc phosphate, iron phosphate, or no phosphate is governed by the end use of the article, and in general follows the same pattern as for conventional painting. In general, zinc phosphate-treated surfaces produce the best corrosion resistance. Our laboratories have found iron phosphates to be suitable in applications where less corrosion resistance is permitted and when one-coat systems are used, the latter primarily because of gloss considerations.

Zinc phosphate conversion coatings which have an extremely fine crystal structure, provide the best substrate for electropaints with respect to surface appearance, adhesion, and corrosion resistance. Although a calcium–zinc process produces an almost amorphous type of coating on steel, it is unsuitable for treating galvanized steel surfaces. The thinner, more densely packed coating produces optimum adhesion and cohesion of the applied organic film since less contaminant in the form of zinc phosphate is introduced into the paint film proper during the electrodeposition process. Zinc phosphate coatings in weight range of 150–200 mg/sq ft offer with many paint systems the best compromise for optimum corrosion resistance and adhesion performance.

Important new phosphating systems for pretreating steel surfaces prior to the electrodeposition of one-coat white paints have been developed. When a steel surface coated with conventional iron, zinc, or calcium–zinc phosphates is electropainted with a white film, the resulting work appears yellow-off white with red stains, blotches, and various types of blemishes showing through the white film. It is believed that the main source of this discoloration is caused by the oxidation of metallic iron at the anode to various types of iron hydrates which become part of the paint film. The iron hydrates entrapped in the paint film are dark and tend to discolor the film. In addition, the hydrates form iron oxides when the paint film is subjected to heat during the curing operation. These iron oxides because of their dark color then show through the white paint film.

If metallic copper is painted white electrophoretically, the copper is also oxidized at the anode to cupric ions. The cupric hydrates which become part of the white paint film are blue and thus discolor the white paint film. For example, in an amine-solubilized system the cupric ions combine with the amines in the paint system to form the intense blue copper ammonium complex, $Cu(NH_3)_4$.

Phosphating systems have been formulated by incorporating small quantities of copper in the coating solutions. The resulting phosphate coatings include traces of copper which during the electrophoretic painting is oxidized to the characteristic blue hydrates, which function to cancel the undesired colors of the iron hydrates or oxides. The result is an unblemished paint film that is a whiter-white color or a blue-tinted white color, depending on the amount of copper that was deposited on the surface. The amount deposited is regulated by controlling the copper content of the system. This pretreatment system in conjunction with white electropaints presently in use should enable the paint industry to use white formulations successfully in everyday production.

Passivation Acidulated Rinse

When zinc phosphated parts enter the electropaint bath either air dry or wet, the use of a passivating rinse containing both hexavalent and trivalent chromium enhances salt spray results. When a dryoff oven is utilized prior to the electropaint tank, a chromate rinse neither helps nor hinders the final corrosion resistance. One may or may not use a chromate rinse if a dryoff oven is used and obtain the same result. However, a chromate rinse is also desired to minimize the rusting of parts during line stops. Furthermore, it has been generally found that if the work is oven dried without a chrome rinse, the results will be better than air-dried metal even when followed by an optimum chromate rinse system. Oven-dried work with or without a chromate final rinse will always produce better corrosion resistance than an air-dried system followed by a chromate rinse.

Non-chromate final rinses have been developed which produce equivalent corrosion resistance to standard hexavalent chromium containing rinses when painted with electrophoretic primers.

In the case of iron phosphate coatings, an after-treatment with a passivating solution containing both hexavalent and trivalent chromium compounds is absolutely necessary for suitable corrosion resistance. The use of a dryoff oven has no effect on the ultimate quality when these amorphous-coated surfaces are electropainted.

A partially reduced chromate after-rinse leads to a reduction in the area occupied by the press. The trivalent and hexavalent chromium rinse actually leads to a "clogging" of the pores by the formation of a chromium chromate gel.

Deionized Water Rinsing

A deionized water rinse is always recommended as the final step in the treatment of the metal, whether or not an oven dryoff is used. This prevents carry-over of foreign ions into the paint bath. It has been shown that when certain concentrations of chloride or chromate are reached in the paint bath, rupturing of the paint occurs and the paint quality falls off drastically. Above 60 ppm of chloride ions the salt spray resistance decreases, and at 100 ppm scribe failure is severe. However, it is not chromates or chlorides *per se* which affect electropainting baths but any electrolyte which might elevate the conductivity of the paint bath.

Deionized water application after the passivating chromate rinse will, in general, not alter the salt spray resistance of an electropaint primer. The time between the application of the chromate rinse and the deionized rinse was not significant when dwell times were varied between 10 and 300 seconds.

The purity of the rinse stages has been found to be related to the wet adhesion problem previously discussed. When a certain paint formulation is marginal in its wet adhesion properties, a small increase in the conductivity of the DI rinse will tend to produce much poorer adhesion of the wet film. Thus, close control of the rinse stages becomes essential when this problem is at hand. The presence of specific ions in the passivating rinse solution also tends to minimize or completely eliminate the problem.

Drying of Phosphate Coatings Prior to Electropainting

With most paint systems tested, oven drying of the zinc phosphate coating before painting in the general range of 150°–205°C for times of 5–10 minutes greatly increases the salt spray performance compared with parts which were painted air dry.

If the oven temperature greatly exceeds 250°—*i.e.*, 250°–300°C for times of 6–10 minutes—corrosion resistance drops appreciably. Recent work on the thermal behavior of zinc phosphate crystals scraped from a steel surface and on the steel surface itself, utilizing thermogravimetric and differential thermal analytical techniques, have shown the exact degree of dehydration of the zinc crystal at various temperatures. The change in the degree of hydration of the zinc phosphate crystal is followed by a change in the corrosion resistance. Figure 1 illustrates the results of thermogravimetric analysis (TGA) and differential thermal analysis (DTA) of zinc phosphate scraped from steel surfaces. Thermogravimetric analysis is a measurement by a thermobalance of the actual weight change of a sample as a function of time or temperature. A sample is weighed before and after heating to a desired temperature for a desired time.

Differential thermal analysis demonstrates calorific phenomena—*i.e.*, disassociation, allotropic changes, and phase changes. DTA and TGA are often used together, as we did in the zinc phosphate study, to demonstrate decomposition. Thus, as the TGA curve showed a weight loss caused by water evolution, the DTA curve showed an endothermic reaction taking place at the decomposition temperature or temperature where water was driven off.

The first two molecules of water of hydration are lost at 180°C while the last traces of water disappear at 320°C. The corresponding DTA curve shows endothermic peaks at 180°C and 320°C also. The increase in corrosion resistance on using a dryoff oven is caused by the loss of two molecules of water of hydration from the hopeite and phosphophyllite, each crystallizing with four molecules of water of crystallization.

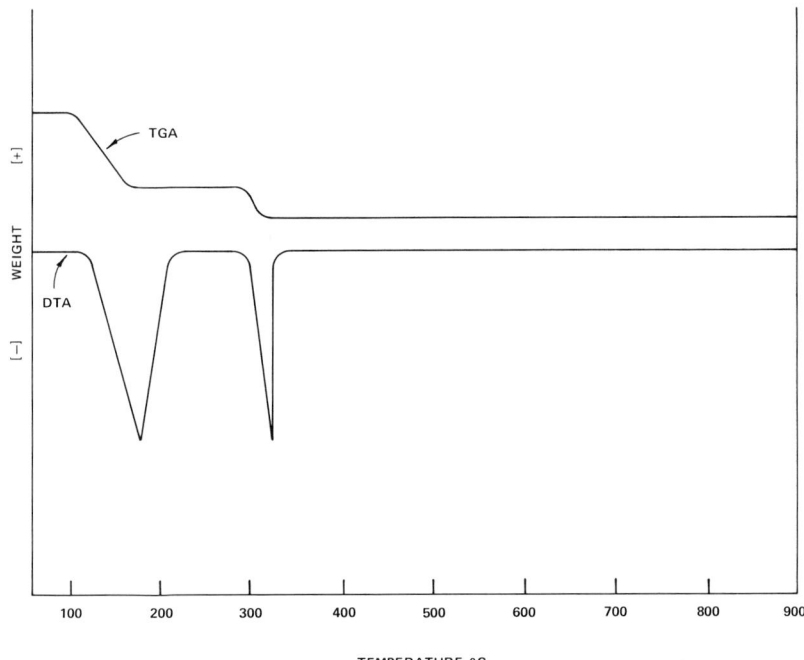

Figure 1. TGA and DTA of zinc phosphate scraped from steel surfaces

Thermogravimetric analysis of a zinc phosphate coating on the surface of steel generally showed the same decomposition temperatures as that of the scraped zinc phosphate crystals. However, above 400°C a weight gain was observed in the TGA curve. As the temperature was gradually increased (heating rate was 10°C/min.), the increase in weight continued at a much greater rate.

Microscopic studies of the crystals were also performed after subjecting the metal to various elevated temperatures. At extremely high temperatures above 430°C the anhydrous crystal actually shrinks to form more voids in the metal surface. Microscopic pictures have shown a shrinkage in the crystal above 400°C. Thus, the TGA analysis in conjunction with photomicrographs indicates that as the zinc phosphated surface is exposed to temperatures above 400°C, the coating shrinks and oxidation of the metal takes over rapidly, causing a sharp increase in weight.

Rinsing after Painting

Some paint systems are sensitive to tap water rinsing even if followed by a final deionized water rinse. Most paint systems, however, can be rinsed with tap water if the final rinse before drying is deionized water. If tap water is used as a final rinse after painting, spotting occurs on the surface either by reaction of the tap water salts with the uncured paint or evaporation of the tap water itself. A trace quantity of wetting agent is added to the final deionized rinse by some paint manufacturers.

Pretreatment of Zinc and Aluminum

The pretreatment of zinc, in general, follows the same rules as the pretreatment of steel. For passivated galvanized steel sheets a zinc phosphate solution containing fluoride will be helpful in forming the coating.

The most important steps in treating aluminum prior to electropainting are alkaline cleaning and deoxidizing the surface. If the surface has scratches and other imperfections, etching the aluminum in strong alkali solutions followed by a desmutting operation is often desirable. Our recent research work has shown that chromate coated aluminum performed better in acid salt spray after electropainting than alkali etched and deoxidized aluminum. The use of acetic acid to lower the pH of salt solutions used in accelerated test programs has shown that it is a useful tool in distinguishing between a good composite system and one which is inferior.

For aluminum parts which are contiguous to steel, and the parts are zinc phosphated, conventional treatments will coat only the steel. Thus, the part will be zinc phosphated steel and bare aluminum. Since there is a large difference in conductivity between zinc phosphated steel and bare aluminum, there is a possibility that rupturing of the electropaint on the more conductive aluminum surface will occur. Therefore, a treatment which will zinc phosphate both aluminum and steel would be recommended in this case.

3

Changes in Zinc Phosphated Steel Surfaces during Electrodeposition

C. A. MAY

Shell Development Co., Emeryville, Calif. 94608

> *During the electrodeposition of a surface coating 30–40% of a phosphatized surface can be electrodissolved because of a decrease in the pH. Interestingly, this loss of the phosphatizing material does not reduce the corrosion protection, and scanning electron microscope studies show that the phosphate crystal coating is not ruptured. Previous studies have shown that the most probable cause for reduced corrosion protection is the electrodissolved ions which are occluded in the coating. Potentiodynamic anodic polarization curves show that the water of hydration in the crystal structure also changes during electrodeposition. This source of reduced corrosion protection, however, depending on the phosphating treatment, can be corrected by heating.*

Most of the vehicles used in the electrodeposition of surface coatings are carboxylated versions of conventional coating molecules. The carboxyl group is required for aqueous solubility by neutralization with various bases, generally organic amines, and the resultant ion provides the necessary electrical charge for attraction to anodic surfaces during the application of the coating.

The chemistry of these anodic depositions is quite complex and has been widely discussed in the literature. Although there is not complete agreement among the various investigators, a number of the anodic events have been well defined. During the coating process the pH at the anode is reduced (*1*) which results in coagulation of the resin on the anode surface. Additional coagulation of the resin results from the cations generated by electrodissolution of the substrate (*2*)—e.g., in the case of a steel surface, ferrous ions. In addition the carboxylate ions can undergo a Kolbe oxidation (*3, 4, 5*). The picture becomes even more complicated considering that most steel substrates are pretreated by some sort of phosphating process prior to coating. Menzer (6) has stated that depending on the nature of the substrate and the electrodeposition process, 40–60% of a phosphated surface can be dissolved during electrodeposition.

Our own studies have shown that the metal ions generated during the electrodeposition process are occluded in the coating rather than passed into the electrodeposition bath (2). We were subsequently able to demonstrate that these occluded ions are responsible for reduced corrosion protection as measured by salt spray testing (7). We also found that if a phosphated steel panel was soaked in water for as little as one hour, allowed to dry, and coated by solvent application, the salt spray protection was also reduced. This latter effect was subsequently traced to an increased water of hydration of the phosphated surface by analogy to the recent work of Chance and France (8). These authors demonstrated that it was possible to predict which phosphated surfaces would give inferior results by the use of anodic polarization scans.

Using a similar polarization technique we studied the effects that the electrodeposition process had on the hydration of phosphated surfaces. This approach proved useful not only in defining the nature of various phosphated pretreatments but also for estimating desirable baking schedules for electrodeposited coatings.

General Experimental Approach

A general discussion of the experimental approach at this point appears advisable for a better understanding of the data which follows. If a metal substrate is placed in an aqueous solution of an inorganic salt and connected to a standard calomel electrode (SCE) through an in-line voltmeter, a potential is noted. In the case of a phosphated steel surface this is around −0.6 volt. If a potentiostat is placed in this system and a voltage is applied in a positive (anodic) direction, electrodissolution takes place and a current is developed. The magnitude of the current is a measure of the rate of dissolution of the substrate at the applied voltage. By increasing the voltage at a uniform rate, a plot of the amperage as a function of the voltage can be obtained. If the rate of voltage increase is kept constant for all experiments, the polarization scans are quite reproducible for a given set of experimental conditions.

Results and Discussion

We decided to examine three types of phosphated steel surfaces using this approach to ascertain how various metal pretreatments may influence the electrodeposition process. The three surfaces used were typical commercial products (Parker Rustproofing Div., Hooker Chem.), A-Bonderite 100, B-Bonderite EP-2, and C-Bonderite EP-89:

A. A standard zinc phosphate process, nitrile accelerated.

B. Zinc phosphate process, nickel and fluoride modified, nitrile accelerated.

C. Zinc phosphate process, calcium modified.

Pretreatment A is a commonly used metal pretreatment for quality finishes. Pretreatment B is a modified version of the zinc phosphate process recommended as a good starting point for the evaluation of electrodeposition coatings. Pretreatment C is recommended for light colored electrocoats.

Table I. Metal Concentration Found in Electrodeposited Coatings on Various Phosphated Surfaces

Pretreatment	A	B	C
Iron, % w	0.07	0.06	0.03
Zinc, % w	0.95	0.57	0.24
Nickel, % w	—	0.45	—
Calcium, ppm	—	34	280

An electrodeposited coating on each type of substrate was first examined for the concentration of the various occluded, electrodissolved metals. The results are given in Table I. The electrocoating vehicle used in this determination and throughout the investigation was a maleinized, oil modified, epoxy ester of the bisphenol type (9). As seen, pretreatments A and B gave approximately the same concentration of metal. The nickel modified pretreatment yielded a lower zinc concentration, but this was compensated for by the occluded nickel ions. The calcium phosphate modification (C) yielded much less metal, particularly electrodissolved iron. The latter fact probably accounts for the better color of electrocoatings on this surface since iron salts discolor during the baking of the finish.

The three phosphate surfaces were first examined as received from the supplier. Prior to sampling, each substrate was heated for three minutes at 190°C to ensure the uniformity of the starting surface. This is an important precaution since the water of hydration will increase slowly on storage under normal laboratory conditions (7). Each pretreatment gives a characteristic polarization curve as shown in Figure 1. The reproducibility of the data is quite good, and hence the curves could be used for identification purposes in much the same manner that infrared spectra are used to identify organic compounds. The dotted portions of the curves are anomalies which sometimes appear in the anodic scans, probably as a result of variations in the phosphate pretreatment over the

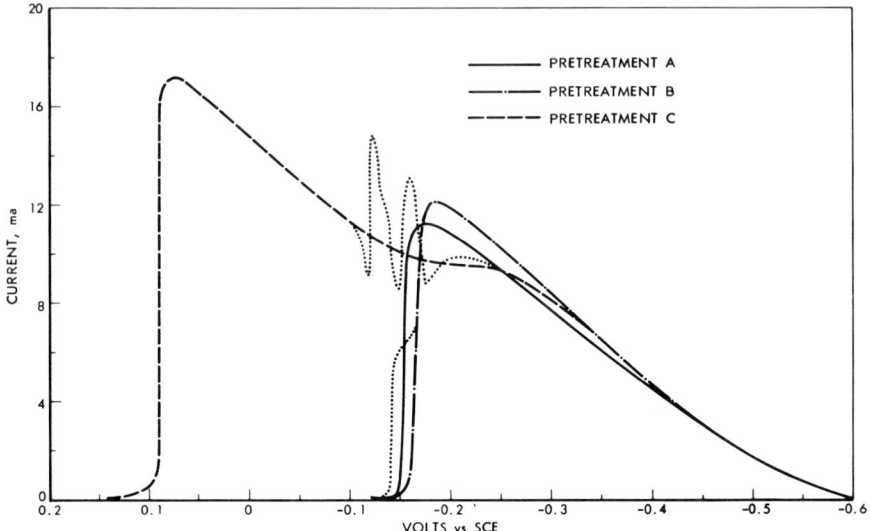

Figure 1. Anodic polarization curves for various phosphated surfaces

surface tested. The standard zinc phosphate (A) and the nickel modified zinc phosphate (B) surfaces give similar results, but the calcium phosphate surface is quite different.

These differences are also seen by microscopic examination of the three substrates. Figure 2 shows scanning electron microscope photomicrographs of each surface at 1000 × (times) magnification. As might be expected pretreatments A and B appear quite similar, but C, the calcium phosphate modification, is quite different. In the latter case the phosphate crystals appear to be much less intimately associated with the steel substrate.

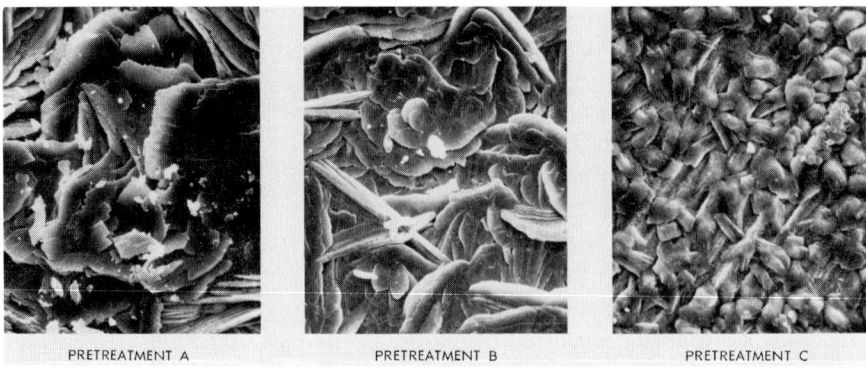

Figure 2. Phosphated surfaces before electrodeposition (× 450)

The effect of electrodeposition (E/D) on the three phosphated surfaces is shown in Figures 3, 4, and 5. The samples were prepared by electrodepositing the epoxy ester on the substrate in question, washing the film off with tetrahydrofuran, drying at room temperature, punching out the test specimen, and determining the anodic polarization characteristics.

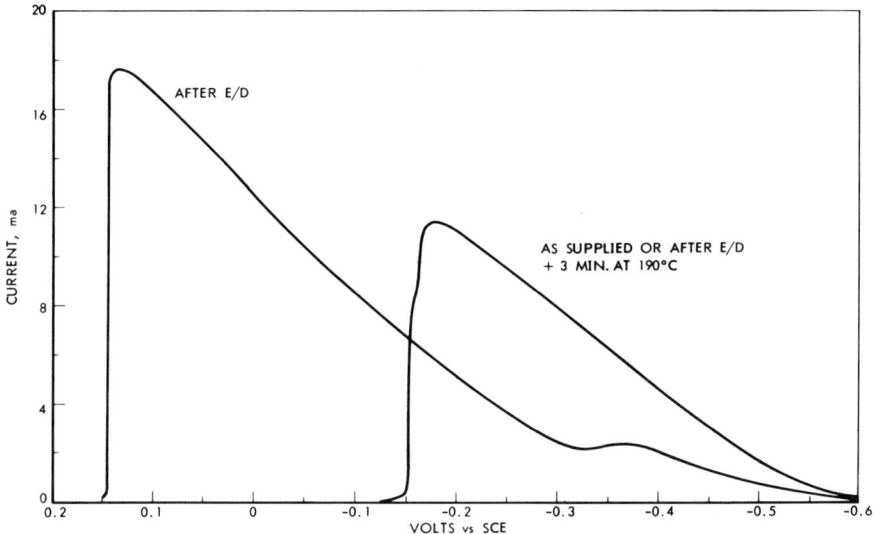

Figure 3. Effect of electrodeposition on pretreatment A

As Figures 3 and 4 show, the standard zinc phosphate pretreatment and the nickel and fluoride modification behave in a similar fashion. After electrodeposition the anodic excursions gave peaks at approximately −0.37 and +0.13 volt in each case. By heating either pretreatment for three minutes at 190°C after it had been through the electrodeposition process, the original polarization scan was restored. With the calcium phosphate modified pretreatment the results are more complex. Before electrodeposition the anodic polarization curve looks quite similar to the other two phosphate coatings after electrodeposition. The coating process changes the curve to a position more nearly similar to the original curves for pretreatments A and B. When this substrate was heated to 190°C after electrodeposition, the initial polarization characteristics were not always restored. Although the results with this substrate were inconsistent, the general shape of the polarization curve following the postelectrodeposition heating is as shown in Figure 5. Thus, there appeared to have been some permanent damage to this pretreatment caused by the electrodeposition process.

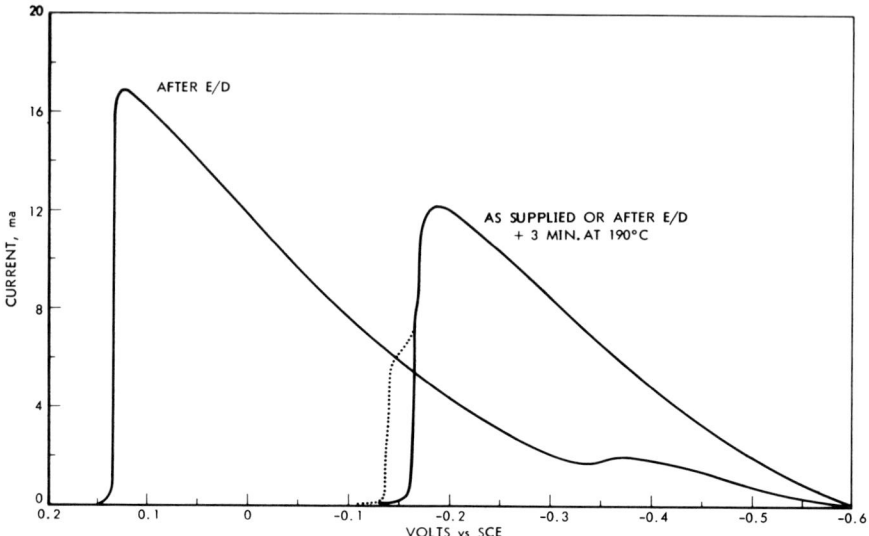

Figure 4. *Effect of electrodeposition on pretreatment B*

As stated earlier, part of the phosphated surface is electrodissolved during coating deposition. The results thus far lead to the conclusion that reducing the amount of phosphating also does not change the polarization characteristics if the substrate is properly heated following elec-

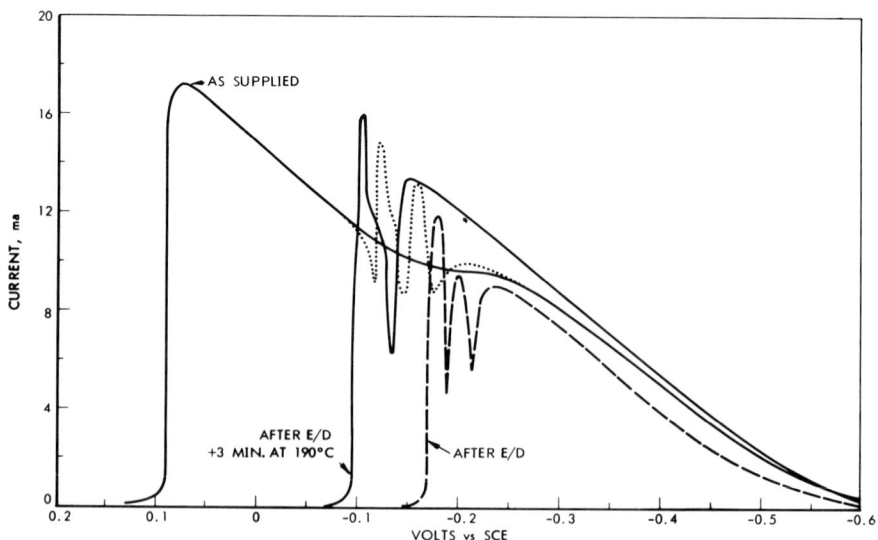

Figure 5. *Effect of electrodeposition on pretreatment C*

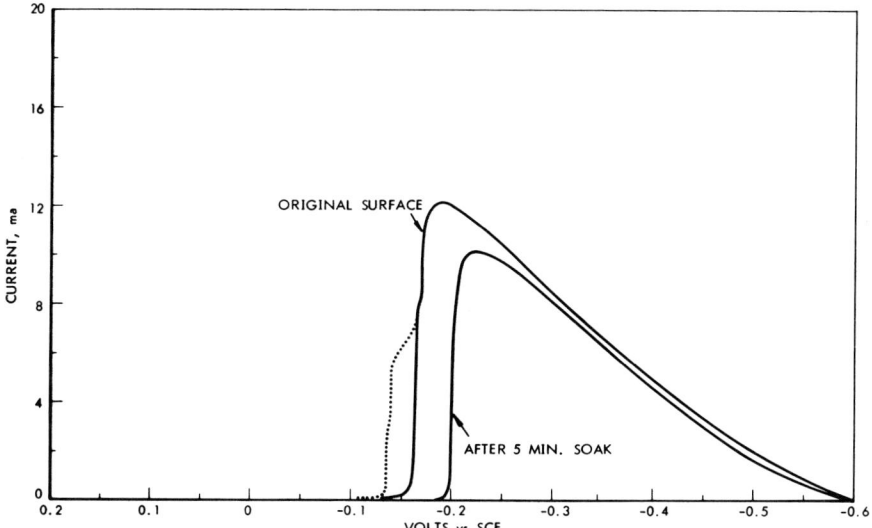

Figure 6. Effect of soaking in E/D bath on pretreatment B

trodeposition and film removal. However, in keeping with the findings of Chance and France, the electrodeposition process or contact with the aqueous bath appears to have increased the water of hydration in the phosphate crystal structure. Investigation of the latter point proved interesting in that it pointed out further differences between the phosphated

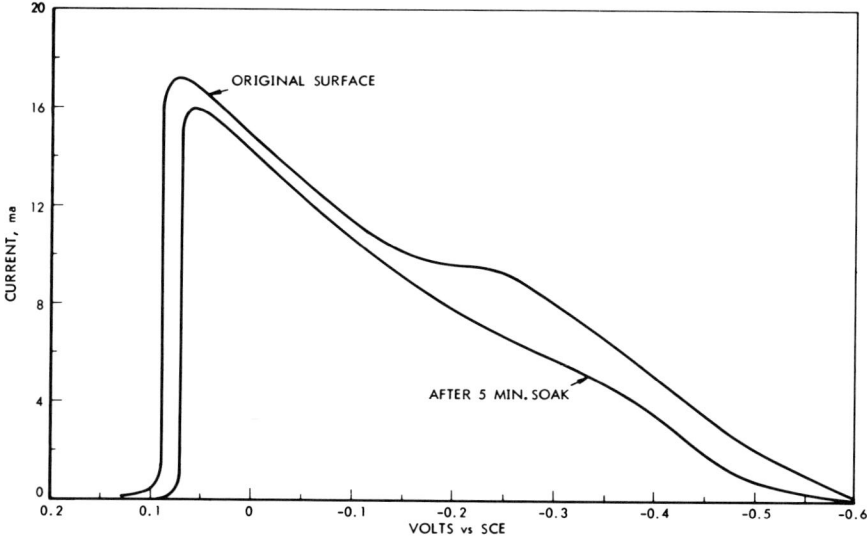

Figure 7. Effect of soaking in E/D bath on pretreatment C

surfaces. During our laboratory operations a maximum of five minutes is required from the time a panel is placed in the electrodeposition bath until the coating is ready to be placed in an oven for baking. Accordingly, panels with each surface pretreatment were soaked in the electrodeposition bath for five minutes, rinsed with tetrahydrofuran, air dried, and examined by anodic polarography. Figures 6 and 7 show that the nickel modified zinc phosphate (B) and the calcium phosphate modification (C) are essentially unaffected by this procedure. The standard zinc phosphate (A), on the other hand (Figure 8), partially approaches the characteristics of a panel which has been subjected to the electrodeposition process. This in a sense explains why the nickel and fluoride modified zinc phosphate may have been the recommended pretreatment for electrocoating. The standard zinc phosphate finish changes somewhat by merely being exposed to the electrodeposition bath; the calcium phosphate modified finish is not changed by soaking, but the electrodeposition process changes the pretreatment to the extent that the damage may be permanent. On the other hand, the nickel modified zinc phosphate changes only during electrodeposition and recovers when reheated. It would thus appear that this finish and the standard zinc phosphate should recover the proper degree of hydration if the electrodeposited coating is baked at a sufficiently high temperature.

Our efforts were next extended to an investigation of the proper baking temperatures for the various pretreatments. A preliminary check

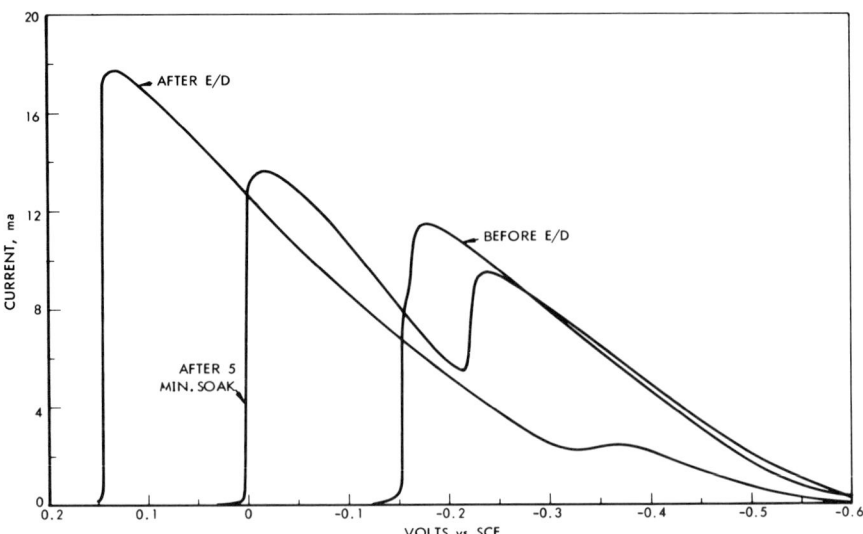

Figure 8. Effect of soaking in E/D bath on pretreatment A

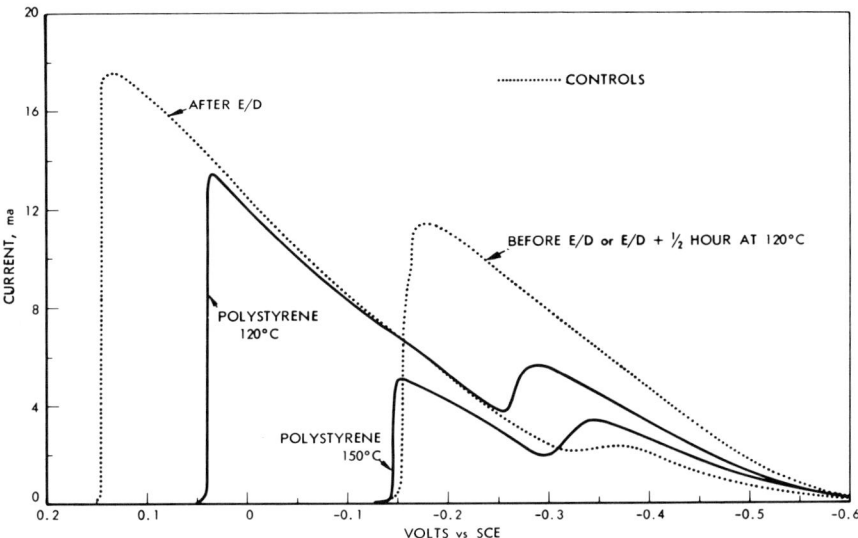

Figure 9. Effect of baking polystyrene film on pre-electrodeposited pretreatment A

showed (*see* Figure 9) that if the uncured, electrodeposited film was solvent washed off of a panel, in this case pretreatment A, heating at the relatively mild conditions of one-half hour at 120°C restored the original polarization characteristics. We suspected, however, that the presence of the coating would retard the "recovery" because of diffusion limitations. As will be shown, this proved to be the case. Because of our suspicions, subsequent experiments were designed as nearly as possible to simulate an electrodeposition procedure.

The first attempt to simulate the electrodeposition process consisted of electrocoating with the maleinized epoxy ester, baking at various temperatures, and then removing the film by swelling in a strong organic solvent. In every case the current peaks of the anodic polarization scans were greatly diminished indicating that some of the coating could not be removed. Accordingly, this approach was abandoned.

The problem was then approached by electrodepositing the coating, washing off the uncured coating with tetrahydrofuran, drying, and immediately applying a coating of polystyrene from toluene solution in a thickness which would be obtained from the electrodeposited coating. The polystyrene coated panel was then baked at the desired temperature, the coating washed off with toluene, and the polarization scan determined. As seen from Figure 9 with pretreatment A a partial "recovery" was obtained after a bake of one-half hour at 120°C. Since complete recovery was obtained under the same conditions in the absence of a coating, the coating definitely slows down the removal of the water of hydration from the phosphate crystal structure. When the polystyrene coating was baked at 150°C, however, difficulties were encountered. Even soaking overnight in the solvent did not remove all of the polystyrene. Apparently, during the baking operation at the higher temperature, the polymer penetrates the phosphate crystal structure to a degree that it cannot be easily removed. The net result is a polarization curve for the 150°C experiment as shown in Figure 9. As judged by the termination of the anodic dissolution at −0.15 volt, the proper degree of hydration has been restored since this is characteristic of the phosphate surface before electrodeposition. However, the double peaks, characteristic of a phosphated surface after electrodeposition, casts some doubt on this conclusion.

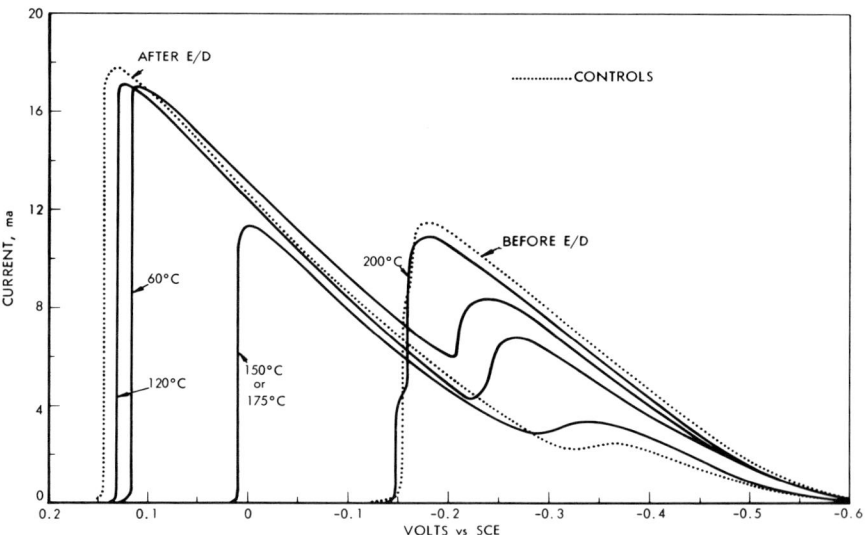

Figure 10. Effect of paraffin heating on pre-electrodeposited pretreatment A

The most satisfactory procedure consisted of electrodeposition, removal of the uncured coating with a solvent, and heating the sample for the desired time and temperature under a pool of molten paraffin wax. Samples thus obtained could be readily cleaned by soaking for a few minutes in benzene. The results showed that this recovery temperature is about 50°C higher than would be required in actual practice where a thin coating is on the surface. However, the data permit an estimation of the minimum desirable baking temperature and also the temperature differences which may be required for proper dehydration of different phosphate treatments.

Shown in Figure 10 are the results obtained with the standard zinc phosphate pretreatment (A). The heating period in each case was one-half hour. At 150° and 175°C a pronounced change occurs in the polarization scan, but complete recovery is not noted until the specimen is heated at 200°C. Combining this data with the results in Figure 9, it would appear that the proper baking for a coating electrodeposited on this substrate is less than 200°C and more likely close to 150°C. Undoubtedly it is much more difficult for the water to escape through the relatively "thick" paraffin layer than 0.5–1 mil of surface coating.

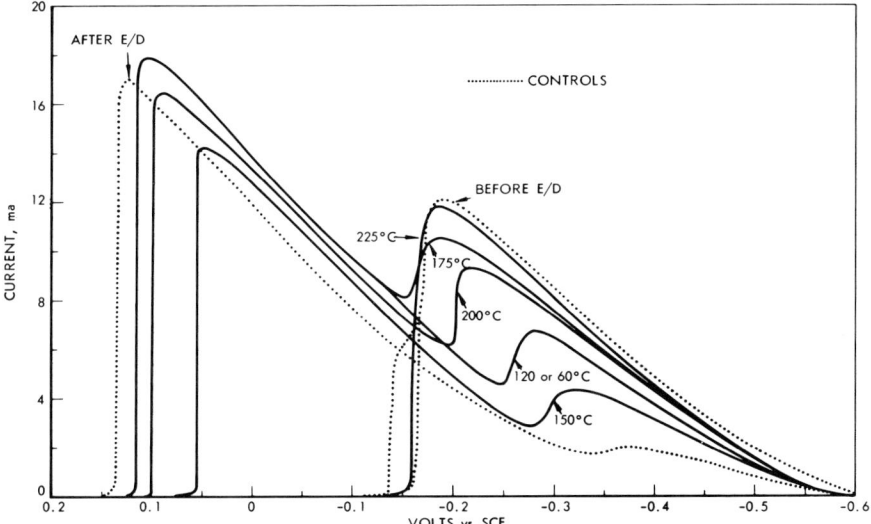

Figure 11. Effect of paraffin heating on pre-electrodeposited pretreatment B

Examination of Figure 11 reveals further differences between the nickel and fluoride modified zinc phosphate (B) and the standard pretreatment (A). Most important is the fact that complete recovery does not occur until 225°C, 25°C higher than that observed with the standard zinc phosphate. Thus if a comparison were made between pretreatments A and B with regard to salt spray performance, it would be expected that coatings baked at 150°C would give similar results, and any differences would only appear when baking temperatures of 175°C or higher were used. The results also show that in addition to the higher recovery temperature required, the partial changes observed at the lower baking temperatures with pretreatment (a) are not as apparent in this case.

The calcium phosphate modification (C) behaved in a markedly different manner from the other two bondcoats (Figure 12). Excluding the aforementioned inconsistencies, which were also evident in these

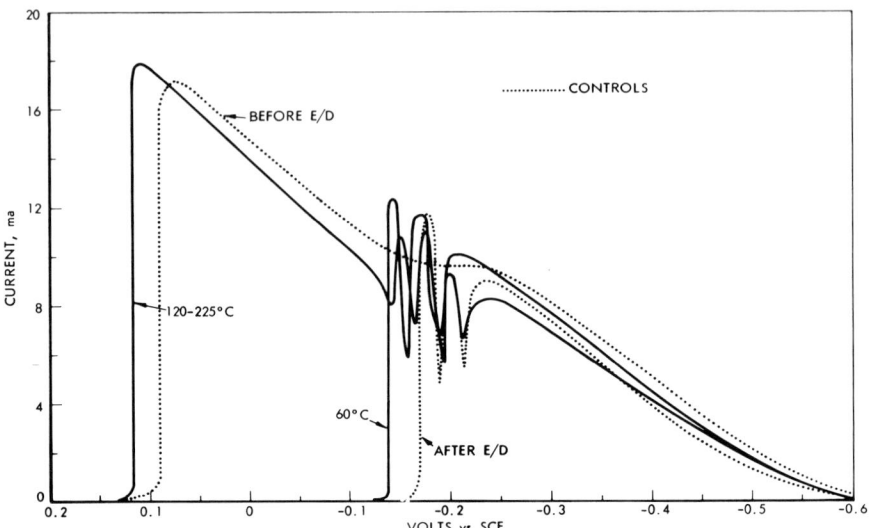

Figure 12. Effect of paraffin heating on pre-electrodeposited pretreatment C

experiments, reversion to the initial polarization scan could occur at any temperature from 120°C upwards.

Visual examination of the three phosphate surfaces after electrodeposition with the scanning electron microscope makes the behavior of pretreatment C more understandable. The photomicrographs are shown in Figure 13. The standard zinc phosphate pretreatment (A) has not changed markedly. Some of the fragmentary crystal structure, however, appears to have been removed. Considering the nickel modification (B), an apparent flaw or crack was discovered at 1000 × magnification. On closer examination (3000 ×), however, even in this area the steel substrate is not visible and the pretreatment appears intact. The calcium phosphate modified pretreatment (C) is a different story. Microscopic examination at 1000 × indicated that there may well be areas of bare (unphosphated) metal showing. This is quite evident at 3000 × magnification. The ridges which are characteristic of the untreated, polished steel substrate are readily apparent. Many such areas can be seen in this picture. The electrodeposition process has actually ruptured the phosphate coating. It thus appears that if the bond coat is ruptured, the anodic polarization scans are erratic, recovery is not always consistent and depends on the area selected for testing. It is also possible that flaws are present in the original calcium phosphate pretreatment and are enlarged by the electrodeposition. Based on the analytical data presented in Table I, since less of the pretreatment is electrodissolved during the

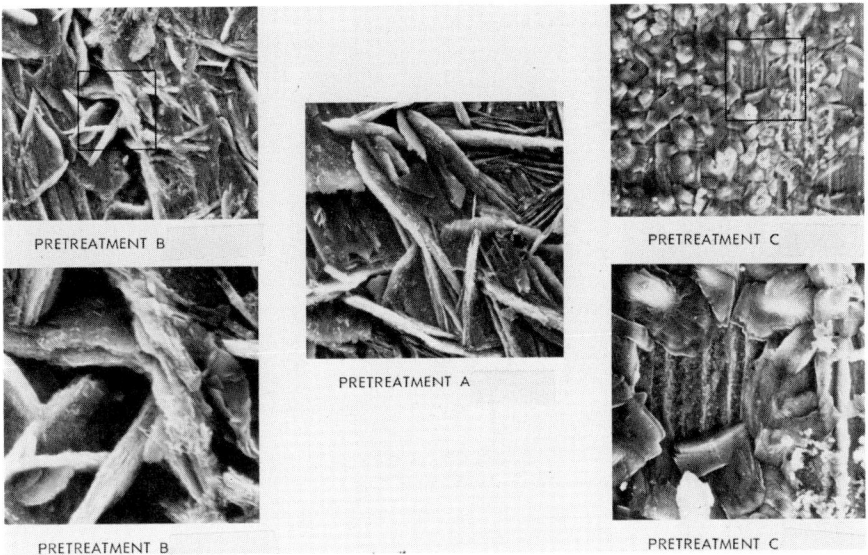

Figure 13. Photomicrographs on the phosphated surfaces after electrodeposition

Magnification: upper left, × 500; lower left, × 1500; middle, × 500; upper right, × 500; lower right, × 1500

coating process, the ease of rupture would indicate a thinner phosphate coating.

Conclusions

Anodic polarization techniques appear to be a useful tool for studying changes in the water of hydration of phosphated surfaces during the electrodeposition of a surface coating. They not only provide a means of identifying various phosphate surfaces but indicate the conditions which change the nature of the surface during the electrodeposition process and how these changes may be rectified.

The nickel and fluoride modification of the phosphating process has the advantage of retarding changes in the phosphate crystal structure resulting from exposure to the aqueous paint system. The changes which are brought about by electrodeposition, however, can only be rectified by baking the coated metal at 25°C higher than that required for a standard zinc phosphate treatment. The important point to emphasize is that evaluations of a pretreatment for corrosion protection, when using the electrodeposition process, should be based on coatings baked over a broad range of temperatures. The differences which are observed may not be the result of a greater degree of cure of the coating at a higher temperature but rather a change in the nature of the phosphate crystal structure.

Experimental

A schematic of the experimental setup is shown in Figure 14. The apparatus consists of a standard calomel reference electrode (SCE), a platinum counter electrode, and the specimen which are connected to a potentiostat. The potentiostat is in turn connected to an X-Y recorder to obtain the voltage and amperage measurements.

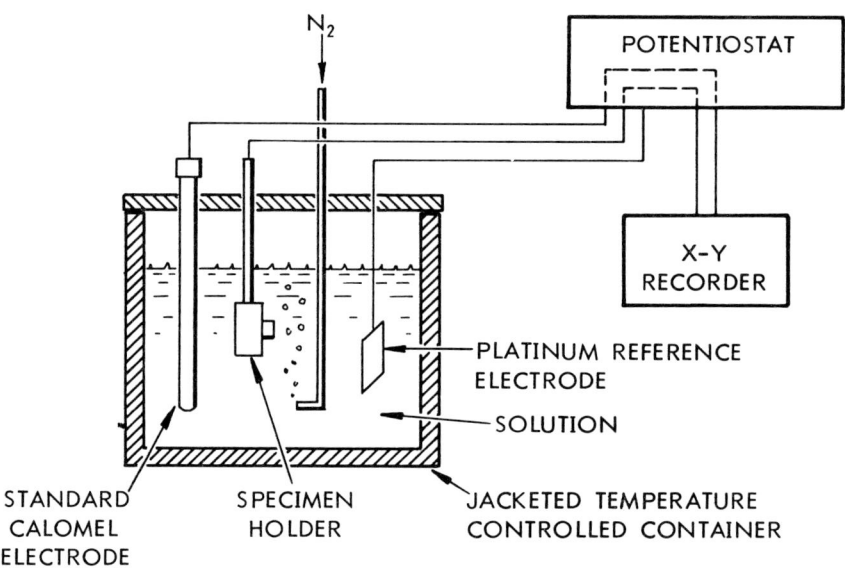

Figure 14. Schematic for anodic polarization equipment

After treating a particular phosphated surface in the desired fashion, specimens were punched from the panel, and one was mounted in a special holder (10). The holder is shown in Figure 15. Microscopic examination showed that the shear punching operation did not damage the phosphate crystals in the test area. Further, the prepared specimens reproducibly retained their polarization characteristics over a period of several days. The time between sample preparation and testing was thus not a factor.

After mounting the specimen in the holder, it was immersed in the electrolyte solution, 0.6M ammonium nitrate. The solution was maintained oxygen free by a high purity nitrogen sparge and thermostatically held at 25.0 ± 0.1°C. Starting from a rest potential of −0.61 volt *vs.* SCE the potential was increased at a rate of 1.2 volts per hour in the

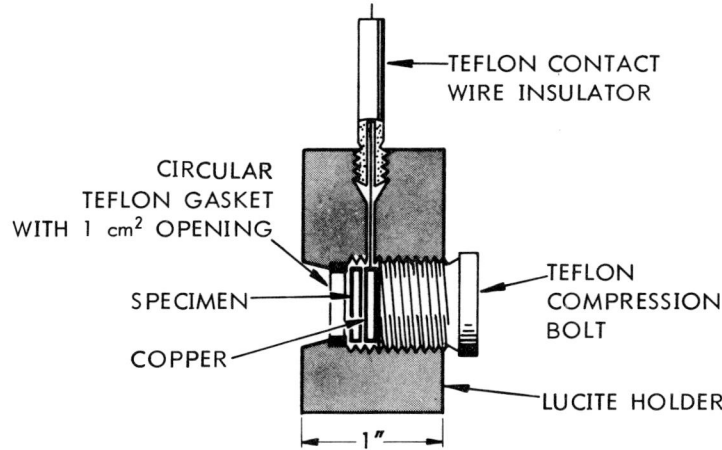

Figure 15. Specimen holder

positive (anodic) direction. The polarization curve was thus obtained by recording the voltage-amperage relationship on the X-Y recorder.

Acknowledgments

The author wishes to thank J. I. Maurer and K. I. Saad for their valuable technical discussions, G. J. McClurg and M. M. Nottage for their part in the experimental work, and R. G. Meisenheimer for the scanning electron photomicrographs.

Literature Cited

1. Nakamura, Y., Komata, K., Nozaki, H., *Bull. Chem. Soc. Japan* (1970) **43**, 663.
2. May, C. A., Smith, G., *J. Paint Technol.* (Nov. 1968) **40** (526), 494.
3. LeBras, L. R., *J. Paint Technol.* (Feb. 1965) **38**, 85.
4. Smith, G., May, C. A., ADVAN. CHEM. SER. (1970) **92**, 140–149.
5. Giboz, J.-P., Lahaye, J., *J. Paint Technol.* (Sept. 1970) **42** (548), 501.
6. Menzer, W., *Product Finishing* (June 1968) **21** (6), 92.
7. May, C. A., *J. Paint Technol.* (Jan. 1971) **43** (552), 43.
8. Chance, R. L., France, W. P. Jr., *Corrosion-Nace* (Aug. 1969) **25** (8), 329.
9. Preliminary Technical Information, RES: 68:15, Epikote Resin Ester DX-31, Shell Chemicals U.K. Ltd., London, SE 1.
10. France, W. D. Jr., *J. Electrochem. Soc.* (1967) **114**, 818.

RECEIVED May 28, 1971.

4

Power Supplies for Electrodeposition of Coatings

CLAYBOURNE MITCHELL, JR.

Elcoat Systems, Inc., 38780 Grand River Rd., Farmington, Mich. 48024

The basic principles of power supply (rectifier) design for electrodeposition of coatings are examined. All system components from the primary ac input to the dc output, along with design criteria, are briefly discussed to aid scientists and production personnel in selecting power systems. Experimental measurements were made of current demand during the electrocoating cycle. These confirmed that the most severe transient conditions on the rectifier occur during the high initial current risetimes of 200–1000 μsec, which rapidly decline 25% or more in 1–3 seconds.

Electrical power for industrial and domestic consumers is supplied in the United States and throughout the world principally as alternating current, and as such, it periodically flows in opposite directions. For industrial processes requiring essentially unidirectional current flow, it is necessary to use appropriate equipment to rectify the current. Consequently, an alternating current (ac) to direct current (dc) conversion is effected, and the equipment is typically called a rectifier, a name which originated during the early era of electroplating.

Basic Components of a Rectifier

This paper discusses the basic components of a typical power supply (rectifier), particularly as related to an electrocoating process. Consideration of its design and functions include the type of output controller, manual or automatic control features, metering, protection, alarms, and power supply ripple. The basic components of a rectifier are shown in Figure 1, and each of these is discussed in detail below.

Ac Input. The incoming industrial power utilized is typically three-phase ac for outputs greater than 5 kw, and the input ac voltages may range from 208 to 550 volts, depending on geographical locale and the age of the plant's electrical system. Thus, where it is economically feasible to operate a low power laboratory supply from a single phase, 120-volt ac circuit, it is quite the contrary for a large manufacturing installation.

Immediately following the input ac terminals, electrical code requires a three-phase disconnect switch or circuit breaker, either internal or external to the rectifier. This makes it possible to isolate safely the incoming power for maintenance. Following the disconnect switch or circuit breaker, there can also be a contactor or starter (which may be remotely operated) to control the input power during various operational phases. A contactor is a magnetically actuated switch to energize and de-energize the primary ac current. A starter is a contactor with thermal overload sensors that automatically disengage the starter in the event of excessive ac current for a period of time. Where a saturable reactor or silicon controlled rectifier (SCR) primary controller is used, it is frequently used as the power deactivator to save the additional expense of a contactor or starter.

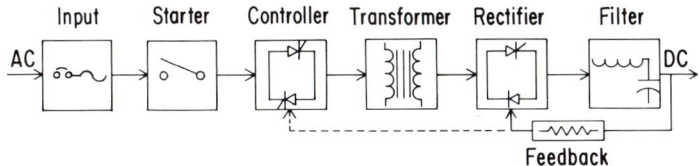

Figure 1. Basic components of a rectifier

Controller. Although the controller may be located in the primary or secondary currents of the power transformer (subsequently discussed), most present day electrocoating rectifier designs utilize them in the primary. The primary currents are typically less than the secondary, and for certain types of solid state control, primary control results in less ripple.

The simplest and least flexible control results from not having one. The output voltage is determined by line voltage conditions ($\pm 10\%$), transformer turns ratio, and load ($-10/15\%$ at full current). For a non-sensitive application, with no regard to coating thickness or quality, coverage can be obtained. However, compensation cannot be made for variables, including the most sensitive—the coating solution.

A tap switch control, manual or motor operated, offers limited range control (30–100%) in a discrete number of steps that varies the output voltage but achieves no regulation in response to line or load changes.

Variable transformer control, manual or motor operated, gives a range (0–100%) of almost stepless output control but otherwise has the regulation deficiencies of a tap switch. Neither a tap switch nor variable transformer is economically feasible for high power plant installations.

The induction regulator is a motor driven (although manually operable for emergency operation) mechanism which "bucks out" or nulls part of the incoming ac voltage to provide continuously variable control of the output voltage. This type of control is infrequently encountered in operating systems owing to the size, cost, and limited control accuracy of a closed loop with a motor drive as one element of the system.

A saturable core reactor control is frequently used for electrocoating rectifier control, and it, plus SCR's, account for probably 90–95% of current production. The operational principle of the reactor is based on the incoming ac current's being looped around a core of magnetically saturable material. A bias winding is also looped through the core, and dc current is applied to this winding. In the absence of a dc bias current, the core material is unsaturated during the ac cycle, and the voltage output is minimum whereas full bias current results in core saturation (minimum inductive reactance) and the maximum voltage output. The output range of control is 5–95% of full output voltage.

Of particular interest with this type of control, as well as the SCR, is its effect upon the incoming sinusoidal voltage, as supplied by the electric utility. Whereas the tap switch, variable transformer, and induction regulator decrease the amplitude of the incoming wave and conduct throughout the entire cycle, a phasing control of the saturable reactor and SCR type do not alter the amplitude but rather conduct for only a fraction of the cycle.

This is shown in Figure 2 where, for example, conduction would occur from 135 to 180 degrees on the first half of the cycle and 315 to 360 degrees on the second half cycle. The equation for the output voltage is:

$$\text{RMS} = \left[\frac{1}{\pi} \int_{\theta_1}^{\pi} E_p^2 \sin^2 \omega t \, dt \right]^{1/2}$$

where E_p = peak voltage
ωt = phase angle of incoming voltage (0-π rad)
θ_1 = angle at initiation of conduction

As a result of this method of varying the output voltage, the incoming voltage and current are chopped into segments. This results in waveform distortion, and the output ripple is greater than that produced with a tap switch or variable transformer controller. Ripple is discussed below.

An SCR is a solid state phasing control which operates on the same principle as a saturable reactor. Its chief differences are a considerably

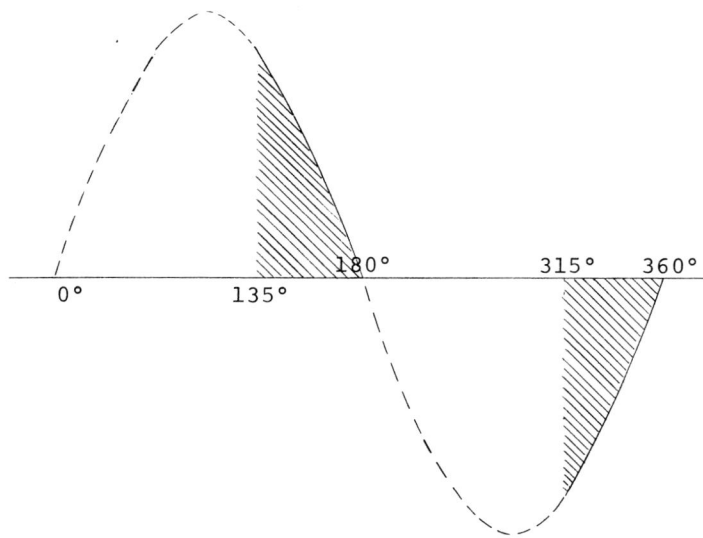

Figure 2. Phase controller conduction angle

smaller size, greater efficiency, faster response, and 0–100% operating range.

Proponents of saturable reactor controllers cite their ability to withstand heavy current surges, which may occur because of large tank loads or shorting of the tank by dangling or scraping parts. However, good engineering design of an SCR controller protects it against these surges, as well as the tank and load.

Manual/Automatic Control. Manual and automatic controls are available for laboratory or production rectifiers. A manual control establishes a nominal output voltage, which may be affected by line voltage or load changes. In an extreme case, the output voltage could drop as low as 25%, for example.

An automatic control samples the output voltage and/or current and compares it with a reference voltage. Differences between the output and reference constitute an error signal which is used, after appropriate signal conditioning, to alter the output appropriately. Typical regulation achieved in practice is ±2% for a saturable reactor or induction regulator and ±$\frac{1}{2}$% or better for an SCR.

Transformer. The transformer serves two primary purposes:

(1) Step-up or step-down of the primary line voltage to the necessary secondary voltage, so that in conjunction with the selected circuit configuration, the required dc voltage is obtained.

(2) Isolation of the load current from the line current, making it possible to ground either polarity of the dc output, or neither.

Rectifier or Rectifier/Control. In the control and/or rectification (shown in Figure 1) which may occur in the secondary circuit of the transformer, the controller is usually in the primary. In the cases where the controller is in the secondary, it could be a saturable core reactor but is more typically SCR's or SCR's and diode rectifiers. With a primary controller, diodes are used in the secondary because only rectification is required.

SCR's in the secondary circuit of the transformer perform the same function as in the primary—phase control of the alternating voltage. This type of control is particularly advantageous if the incoming primary voltage is greater than 600 volts ac—for example, 2400 or 4160 volts. A primary SCR controller would require a step-down isolation transformer to reduce the incoming voltage to a value (220–550 volts ac) compatible with the SCR peak reverse voltage rating. A secondary SCR is subjected to the potentials typical of electrocoating voltages and therefore does not require additional transformation.

Diodes are combined with SCR's for cost reduction (for comparable current and voltage rating, diodes are cheaper than SCR's) and simpler SCR triggering requirements. However, where the current output requires the paralleling of SCR's, careful engineering is necessary to assure balancing of the SCR currents—a more difficult task than balancing only diode currents. In addition, the SCR–diode combination produces a greater ripple output than a primary SCR or all SCR secondary controller. This necessitates greatly increased filtering requirements where the specifications indicate a low ripple (10–20%) at short conduction cycles (25–30% of full voltage output).

Operation of a Rectifier

Circuit Protection. Circuit protection appears in various forms and serves the multiple purposes of personnel and equipment protection, minimization of fire hazard, and protection of the product being electrocoated. Fuses may be used to protect SCR's, diodes, transformers, and other components, often isolating the faulty part or simultaneously shutting the rectifier off. Circuit breakers are used for the same purpose and require only re-engagement rather than replacement to resume operation.

In the event of inadequate cooling of SCR's, diodes, transformers, or other components, temperature sensors are used which protect by shutting off a part of or the entire system.

Electrocoating rectifiers operate in industrial plants in which there may also be welders, large motors, or other electrical machinery which can cause line voltage transients. These transients may be as large as several hundred volts and may cause damage to SCR's, diodes, and other solid state devices, unless properly protected. Protection usually consists of snubber networks—resistor/capacitor combinations which bypass the transient around the protected component; selenium surge suppressors which absorb the transient and dissipate it as heat, zener diodes, or gas discharge devices.

In an electrocoating operation, tank shorts can occur because the part (anode) is oversized or sufficiently agitated by the solution movement to touch the tank wall (cathode). Similarly, in a conveyor operation, a part might drop and become wedged. Consider, for example, the effect of a car body which, while remaining connected to the anode, becomes jammed against the tank. The several possibilities which could occur include burning a hole in the tank, melting part of the body, or drawing such excessive current that the rectifier is destroyed. This type of protection is discussed below.

Finally, some degree of protection is offered by audible and/or visual warning systems in which it is desired to indicate a malfunction without ceasing operation. However, these should be supported by failsafe measures in the event the warning signals are unnoticed or unheeded.

Output Control. The output control can range in complexity from the simple manual setting of a tap switch or variable transformer control to an automatic voltage/current control that will regulate to less than $\pm 1\%$ in voltage and current with 0–100% voltage and current control range. This is accomplished by the sampling, feedback, error signal generation technique previously noted.

The output voltage and current are typically read on $\pm 2\%$ meters, unless more expensive and accurate meters are specified by the customer. Despite the seeming paradox of a better than $\pm 1\%$ control and $\pm 2\%$ meters, this indicates repeatability (as determined by instruments of greater accuracy) of the output setting over an extended period of time or when switched repeatedly off and on whereas the output can only be set within $\pm 2\%$.

Ampere-hour meters are often specified in electrocoating installations to measure the total integrated amount of energy expended for a coating application. The system can be designed so that not only are the total ampere-hours recorded, but a signal can be provided to replenish solids or other constituents as they are depleted.

A control that is commonly specified in conveyorized electrocoating installations is the high–low voltage control. If the conveyor is stopped for many minutes and parts are left in the tank at full coating voltage, excessive film thickness results. Not only is this expensive but often the coating obtained is undesirable, and the part must be stripped and recoated. Turning the rectifier off immediately allows dissolution of the film already deposited, and could result in start-up with a tank full of nearly uncoated parts, thus creating an abnormal current demand on the rectifier. To avoid this problem, the rectifier is programmed so that when the conveyor is stopped, the voltage is switched to a lower value—approximately 50 volts. It may remain at this lower value, or a timer may turn the unit off if the conveyor remains stationary for longer than a given time, typically five minutes.

Rectifiers with automatic controls are typically designed to sense the output current and regulate the output such that it cannot exceed the maximum of which the rectifier is capable. This is achieved in phasing type controls by decreasing the output voltage to a value such that only maximum current is produced. Thus, in the event of a fixed short in the tank or at the rectifier output, the meters might indicate that maximum current was being produced at zero voltage. This is not possible since Ohm's law is still applicable, but a low resistance short requires so little voltage for full output current that the voltage appears to be nil, especially on a 300- or 500-volt scale.

The dc overload is a current-limiting device which also senses a current maximum and either greatly reduces the output or shuts it off entirely. It is normally set slightly higher (5–10%) than the automatic current limit, and as a safeguard to it. However, it is possible to set it lower than the current limit point to guard against overloading the rectifier capability. For example, if too many parts are in an overloaded coating tank, this forces the rectifier to limit current automatically. This reduces the coating voltage, and inadequate film coverage can result, which may not be discovered until in the field. With the dc overload set lower (5–10%) than current, the unit is shut off and the conveyor overloading is immediately detected.

Various claims have been made for pulsed power in electrocoating applications, and it can be programmed into rectifiers that are designed for this application. However, its efficacy has not been sufficiently established to warrant the additional costs involved (25–100%).

If the electrocoating application is batch rather than conveyor, the initial current demand must be considered when power is first applied. Either the rectifier must have a current capability to deliver the first instant of current demand, or appropriate current limiting must be included to protect the rectifier.

Input Ac Requirements. The power factor ($\leqslant 1.0$) of a rectifier is the ratio of the power input (kilowatt) to the kilovolt-ampere (kva) input, which is proportional to the product of root-mean-square (rms) line voltage and current. The power factor is 0.90–0.92 at full output but drops significantly at reduced voltage and current owing to changing load conditions and the waveform distortion created by a phasing type controller.

The efficiency (≤ 1.0) of a rectifier is the ratio of the power output (kw) to the power input (kw). An electrocoating rectifier has an efficiency of 0.92–0.95 at full output voltage and current, which decreases somewhat with lowered output voltages. The decrease results from fairly constant power losses in components such as the wiring and diodes whereas the load power consumption becomes less and the ratio is lowered.

Output Dc Requirements. The output voltage required is principally a function of the electrocoating material, required throw-power (ability to coat deep recesses, etc.), coating time, and tank parameters. The current required is mainly determined by the ampere-hour/pound (or coulomb/gram) specification of the material, the number of pounds per unit area for a given coating thickness, and the area of metal in the tank to be coated. Consideration must also be given to the typical current *vs.* time characteristic of the coating process.

Power Supply Ripple. Although a rectifier output is referred to as direct current, it is not in any sense as direct and smooth a current as that obtained from a battery. It can be smoothed with ripple filtering, typically a combination of inductance and capacitance, to almost any extent, but the cost increases proportionately.

The variations in the dc output result from the three phases of alternating currents flowing in the primary and which may be distorted by the controller. The resultant rectified output represents the overlapping of the tops and inverted bottoms of the transformed primary voltages, and the ensuing crests and valleys. These constitute the ripple variations in the dc output, which are measured with a "true" rms measuring instrument. The "true" results from the inability of ordinary ac panel and portable meters to read the correct rms value of the alternating component. The form (ripple) factor is the rms alternating component divided by the average dc value (this can be ascertained with any average reading meter).

Tap switch and variable transformer controlled rectifiers have a ripple of 4.5–5.0% throughout the entire range of control. Saturable reactors and SCR's have a full output value of 4.5–5.0%, which increases to 50–100% at low output voltages (10–25%). A typically specified value of ripple correction is 10% ripple at 10% voltage and current.

Opinion in the electrocoating industry varies as to the necessity or degree of ripple filtering required. There is some evidence that ripple contributes to increased heating effects in the deposited film, and there is suspicion that coating rupture may be related to ripple crests. Part of the difficulty arises from differences in ripple output of laboratory and production rectifiers, and this contributes to variation in the corresponding results.

Practical Considerations

For those not familiar with rectifiers, yet involved in their selection, a few guidelines that can be used are:

(1) The rectifier should be designed with regard for the safety of operating personnel, the parts being coated, and the rectifier. Proper design allows for access to all parts of the unit but interlocks shut off all power if entry is attempted while energized. The output current should be limited to the maximum of which the unit is safely capable, and provision should be made to shut down in the event of a tank short. The overall design should conform to NEMA and JIC electrical standards.

(2) The design and construction should incorporate components of industrial quality and good engineering practices to maximize reliability. Maintenance must be possible with regular plant personnel, and all parts of the unit must be capable of rapid replacement in the event of a failure. Maintenance should preferably be on a subsystem basis, such as printed circuit boards, or a replaceable chassis, so that service personnel are not required to test components or to be intimately familiar with the circuitry.

(3) The rectifier must include appropriate features that emphasize economical operation, provide flexible operation, and have a minimum of maintenance requirements and minimum obsolescence.

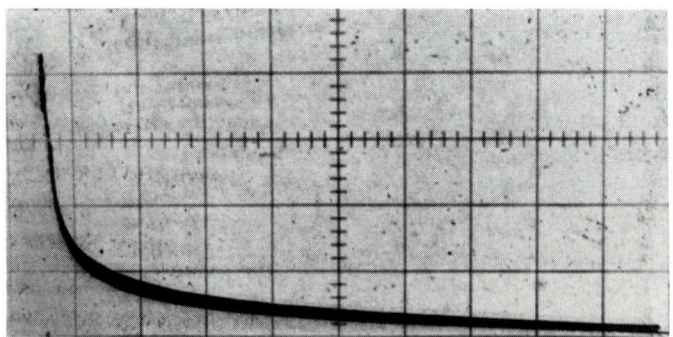

Figure 3. *Current vs. time (HT-300). Horizontal: 3 sec/div; vertical: 0.8 amp/div.*

Electrocoating Current Measurements

The current as a function of time in an electrocoating cycle is not constant, as in a plating bath, for example, but rather exhibits a large initial demand which decreases as the deposited layer accumulates. In this respect the cycle is similar to that of anodizing. The current can be observed with an oscilloscope that vertically displays the voltage drop across a calibrated resistor that is in series with the current flow while sweeping horizontally. The oscilloscope sweep is set to be initiated near the start of the coating cycle. A pen recorder, which has a much slower response than an oscilloscope, will correctly display this curve 5–10 seconds after the initial surge but will not display the rapid current transition at the beginning.

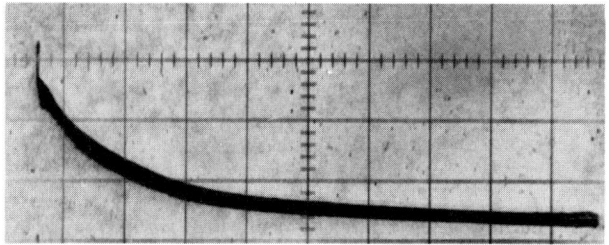

Figure 4. Current vs. time (LT-300 v). Horiz: 3 sec/div; vert: 0.8 amp/div.

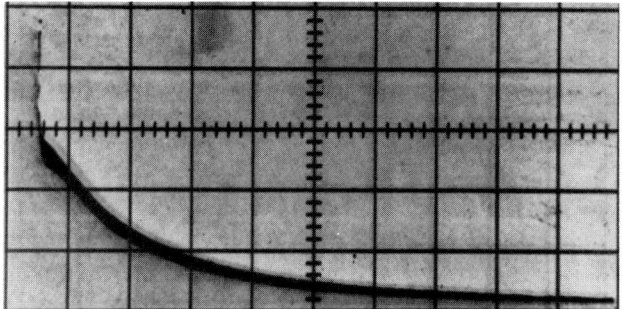

Figure 5. Current vs. time (15303-300 v). Horiz: 3 sec/div; vert: 1.6 amp/div.

Laboratory experiments were conducted to demonstrate the oscillographic display of electrocoating current vs. time, particularly of the initial surge. Measurements were made with Parker Co. Bonderite EP-1 treated panels in a solution volume of approximately 1.7 liters, a chrome-plated steel cathode with an area of 90 sq cm, an anode area of approximately 100 sq cm (one side), and an anode–cathode spacing of 10 cm. The solution temperature was 24°C. Coatings used were Sherwin Williams KS30A (high throw), KS30B (low throw), and PPG S15303 (high-medium throw), and in all cases a potential of 300 volts dc was used. The results and experimental arrangement are shown in Figures 3–11.

In Figure 3 the high throw KS30A exhibited a peak current of 3.4 amp, which dropped to half the initial current in about 0.5 sec. In Figure 4 the low throw KS30B had an initial peak current of 2.7 amp and dropped to half this value in about 3 sec. Figure 5 shows that the S15303 had an initial current peak of 7.5 amp which dropped to half this amount in approximately 1 sec.

The results in Figure 6 were obtained under the same conditions as those in Figure 5 except that the horizontal was changed to 0.5 sec/division rather than 3 sec/division. In this case the initial peak current is 8 amp, with a second peak which occurs about 1.2 seconds after the first. The second peak was later established as being the current surge to the rear of the anode with respect to the cathode.

Figure 7 is a duplication of Figure 3 except that the horizontal was changed to 2 msec/division to demonstrate better the initial surge. The

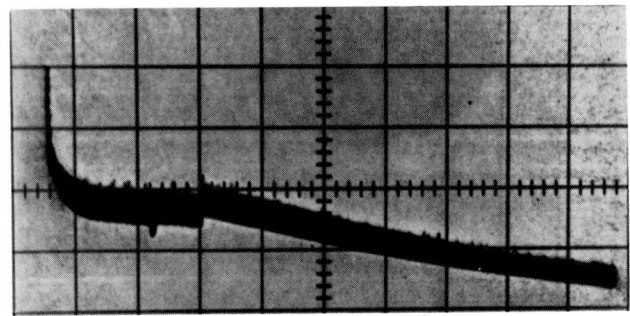

Figure 6. Current vs. time (15303-300 v). Horiz: 0.5 sec/div; vert: 1.6 amp/div.

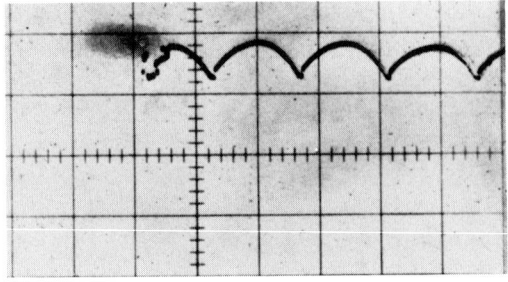

Figure 7. Current vs. time (HT-300 v). Horiz: 2 msec/div; vert: 0.8 amp/div.

power supply ripple (~4.5%) is clearly shown, and this accounts for the width of the current trace shown in the previous figures. The slight distortion of the first cycle of ripple is caused by chatter of the relay used to apply dc potential. However, despite the increased horizontal magnification, it is difficult to estimate the surge risetime, but it is probably on the order of 200 μsec.

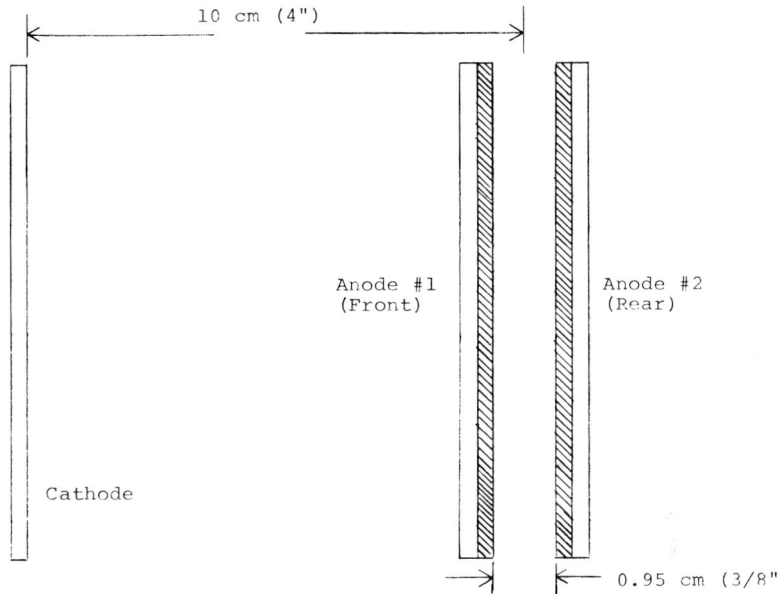

Figure 8. Anode placement for outside front and rear measurements

Figure 8 shows the experimental arrangement used to obtain the data of Figure 9. The anodes consisted of two panels with their interior surfaces masked. Separate anode leads (common cathode) were connected to the power supply, and the anode currents were monitored separately. Anode surface 1 (front), nearest the cathode, is referenced to the abscissa line which is on the bottom; anode surface 2 (rear), furthest from the cathode, is referenced to the abscissa line which is second from the bottom in Figure 9. The peak current on anode 1 is 4 amp, and on anode 2 it is 0.8 amp; both are approximately 0.8 amp within 1.5–2 sec after power is initiated. The peak for anode 2 occurs about 0.5 sec after that for anode 1.

Figure 10 is a sketch of the experimental arrangement for obtaining the data shown in Figure 11. The anodes again consisted of two panels, but the exterior surfaces were masked. Separate leads enabled separate monitoring as before, with anodes 1 (front) and 2 (rear) referenced as

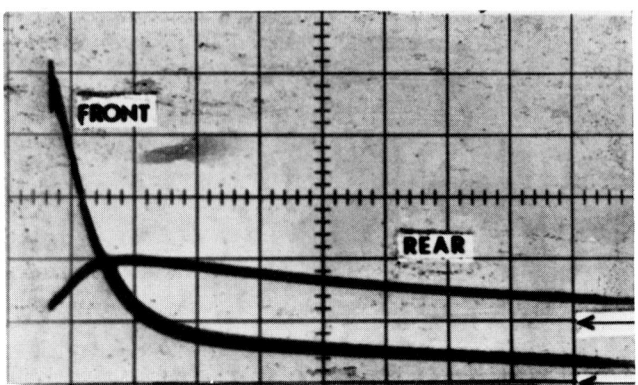

Figure 9. Current vs. time (HT-300 v) outside front and rear. Horiz: 0.5 sec/div; vert: 0.8 amp/div.

in Figure 9. Anode 1 exhibits a peak current of 1.5 amp and decreases to half this amount in 0.2 sec. Anode 2 peaks to 0.8 amp 0.15 sec after initiation of power and decreases to half this value in 0.95 sec.

Figure 12 is a graph showing the current cycle for a phosphated metallic area of *ca.* 1000 sq ft which is entering a large electrocoating tank. The voltage remains constant at about 170 volts throughout the 45-second period. The current demand is initially 1350 amp, decreases

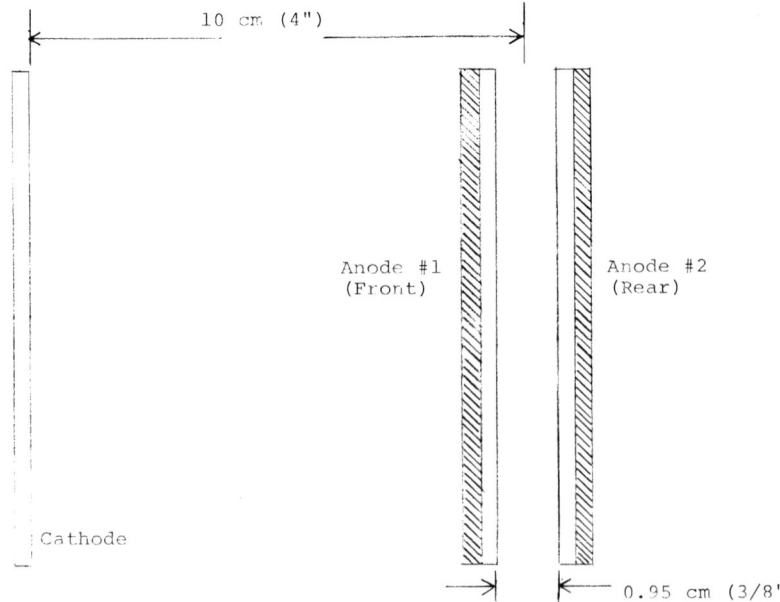

Figure 10. Anode placement for inside front and rear measurement

to 1000 amp in 0.3 sec, and is down to 800 amp in 4.3 sec, gradually decreasing thereafter.

Figure 13 has the same vertical scale as Figure 12, but the horizontal is changed to 0.2 sec/division. The faster horizontal sweep exhibits less of the total cycle but more of the initial rise. Even at this sweep rate, the risetime of the initial pulse is quite short. Laboratory measurements of samples in small tanks indicate a risetime of 200–1000 μsec.

A curve of this type can be integrated to yield the total number of coulombs, and the weight of the sample will give the total grams of deposited material to determine the coulomb/gram ratio. However, an ampere-hour meter or a chart recorder will give reasonable accuracy because the initial surge exists for such a short time. An ampere-hour meter is completely acceptable, for example, in conjunction with an electrocoating rectifier for integrated measurement of the current to restore depleted solids automatically on a periodic basis.

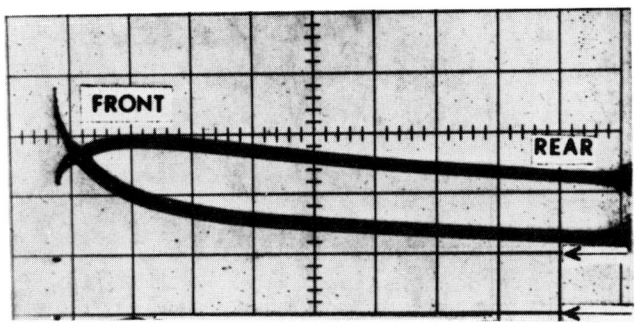

Figure 11. Current vs. time (HT-300 v) inside front and rear. Horiz: 0.1 sec/div; vert: 0.4 amp/div.

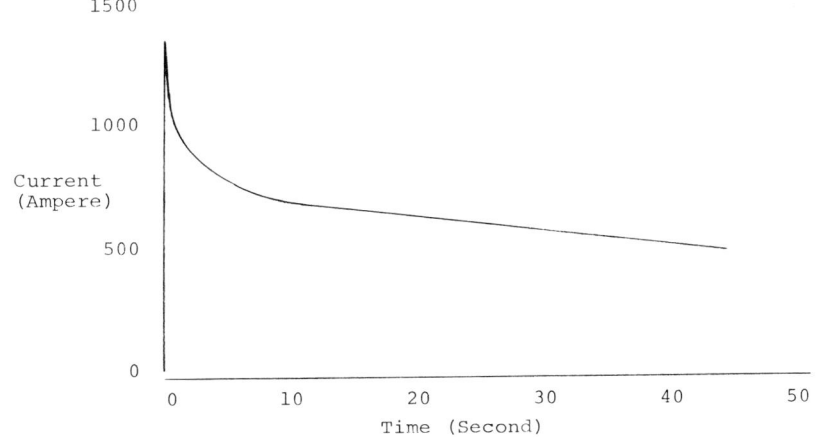

Figure 12. Current vs. time during electrocoat of 1000-sq ft area

Since the data from the laboratory experiments are similar to that obtained from a production facility (especially initial surge), it is concluded that the laboratory data can be useful in designing a power supply for electrocoating applications. Further, the most rigorous duty requirement, in terms of current demand, occurs during the initial coating period. However, this interval is short compared with the total coating time, and this represents a transient condition more than a steady-state one.

The initial current surge can be predicted because it represents the resistance of the coating material between the anode and cathode, as determined by the specific resistance of the material, electrode areas, electrode spacing, and applied potential. This is more easily determined for an experimental arrangement than for a complicated shape in large production.

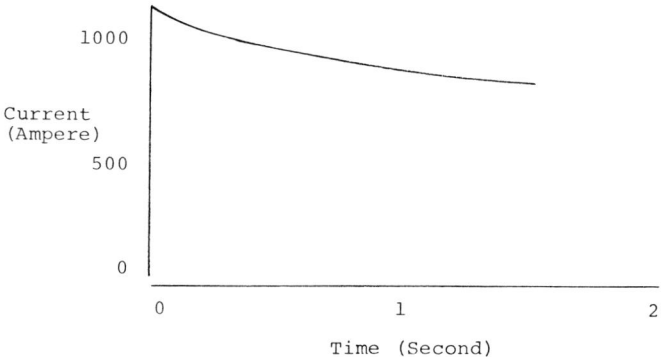

Figure 13. *Initial current surge during electrocoat of 1000-sq ft area*

Finally, note that the initial surge of 1350 amp shown in Figure 12 is equivalent to an incoming primary ac current of 20–25 amp/μsec (neglecting circuit resistance). Consequently, for the power supply circuit designer, components such as SCR's must and can be appropriately selected and protected to ensure that reliable operation is obtained under these transient conditions.

Summary

The information presented here on rectifiers will hopefully give those better acquainted with other aspects of electrocoating an insight into an important component of this system. This information will enable the prospective purchaser of such equipment to ask knowledgeable questions and to understand some of the tradeoffs that must be considered, in conjunction with the manufacturer, to realize the most economical performance, reliability, and safety.

Acknowledgment

The author acknowledges helpful discussions with M. Koltuniak, Vice-president of Engineering.

Bibliography

Sandford, J. E., "Electrocoating: The Charge is On," *Iron Age* (Nov. 2, 1967).
DeVittorio, J. M., "The Electro-Mechanical Aspects of System Design for the Electrodeposition of Water Borne Coatings," The Sherwin-Williams Co.
Streeter, K. L., "DC Power Supplies for Electropainting," 1st Conference on Electropainting for the Seventies, Westinghouse Brake & Signal Co., Ltd.
Koch, R. L., "High Gloss White Electropaint," 1st Conference on Electropainting for the Seventies.
Yeates, R. L., "Electropainting," Draper, Teddington, England, 1966.
Burnside, G. L., Brewer, G. E. F., Electrophoretic Coating Process, U.S. Patent **3,200,057** (1965).
Oster, T. H., Cyclical Current Reversal for an Electrophoretic Deposition, U.S. Patent **3,200,058** (1965).
"Canada Leads the World in Electrocoat Efficiency," *Modern Finishing Methods* (1969).
Mitchell, C., "DC Control Systems for Electrocoating Applications," ASTME paper FC69-127 (1969).
Finn, S. R., Hasnip, J. A., "Electrodeposition: A Current-Time Relationship," *J. Oil Colour Chemists Assoc.* (1965) 48.
Cooke, B. A., Strivens, T. A., "The Non-ohmic Nature of Conduction in the Electrodeposition of Paint Films," *J. Oil Colour Chemists Assoc.* (1968) 51.
Ashby, D., "Rectifiers to Provide Current for Electrodeposition Process," *Trans. Inst. Metal Finishing* (1964) 41.
Burnside, G. L., Brewer, G. E. F., Strosberg, G. G., Igras, R. A., "Prediction of Current Requirements (Ford process)," *J. Paint Technol.* (1966) **38** (493).
LeBras, L. R., "Electrodeposition Theory and Mechanisms," *J. Paint Technol.* (1966) **38** (493).
Berry, J. R., "Electrodeposition of Paint—I," *Paint Technol.* (1963) **27** (12).
Berry, J. R., "Electrodeposition of Paint—II," *Paint Technol.* (1964) **28** (1).
Gloyer, S. W., Hart, P. P., Cutforth, R. E., "Electrodeposition: Theory and Practice," *Official Digest* (Feb. 1965).
Tawn, A. R. H., Berry, J. R., "The Electrodeposition of Paint: Some Basic Studies," *J. Oil Colour Chemists Assoc.* (Sept. 1965) **48**.

RECEIVED May 28, 1971.

Polymers and Pigments: Introduction

GEORGE E. F. BREWER

The success of electrodeposition is the result of the development of film-forming organic macro-ions, which may be symbolized as R^+ and $R^{(+)n}$ or $R^{(-)}$ and $R^{(-)n}$. The molecular weight of the organic radical R is usually between 5,000 and 50,000 while electrodeposition indicates equivalent weights of about 1,000 to 2,000.

In most cases the resinous macro-ions can be functionally described as $RCOO^-$ or as RNH_3^+ and are synthesized in paint plants as acidic resins, RCOOH, or alkaline resins, RNH_2. Subsequently, aqueous dispersions are formed by the addition of solubilizers—namely, bases or acids, respectively.

Deposition	Resin	+ Solubilizer	Film-forming Macro-ion	+ Counterion
Anodic	RCOOH	+ YOH aq	$RCOO^-$	+ Y^+ + aq
Cathodic	RNH_2	+ HA aq	RNH_3	+ A^- + aq

Solubilized resins, such as those symbolized above, are used for electrodeposition.

Conventional spray and dip coating materials contain resins and pigments dispersed in liquids which are subsequently evaporated to leave a polymerizable film. Electrocoating baths contain similarly dispersed materials which are deposited on the electrode, virtually free from solvents or water, ready to be polymerized. Thus, the final film properties are largely predictable from the experience gained with spray or dip painting resins carrying similar chemical linkages such as epoxy, acrylic, and others. A variety of resins are successfully applied by the electrodeposition process. Their compositions are reported in the patent literature which has been compiled (1). Special test procedures for the composition of the coating bath and its deposition characteristics have been developed (2).

Electrodeposited cured coats compete with conventional spray or dip coats. The final coats are, therefore, subjected to the conventional tests, such as salt spray, outdoor exposure, alkali resistance, etc.

Literature Cited

1. Ranney, M. W., "Electrodeposition and Radiation Curing," Noyes Data Corp., Park Ridge, N. J., 07656, 1970.
2. Brewer, G. E. F., Hamilton, R. D., "Ford Motor Company's Electrocoating Process: Testing of Materials," "Preprints," *Div. Org. Coatings Plastics Chem.* ACS (Sept. 1967) **27** (2), 184–191; *J. Paint Technol.* (1970) **41** (541), 119–130.

5

A Low Molecular Weight Esterified Copolymer of Styrene and Maleic Anhydride For Electrodeposition

JOHN F. MOTIER and DONALD L. MARION

ARCO Chemical Co., Division of the Atlantic Richfield Co., Philadelphia, Pa. 19145

> *A partially esterified low molecular weight copolymer of styrene and maleic anhydride has demonstrated exceptional properties as an electrodeposition vehicle. When crosslinked with an amino resin and cured at 360°F for 25 minutes, the copolymer yielded electrodeposited films (unpigmented) possessing salt spray resistance in excess of 450 hours. Variations in curing cycle had little effect on corrosion resistance. Throwing power (steel pipe method) was in the range of 7–8 1/2 inches, and the coated throwing power strips exhibited excellent corrosion resistance. By employing a baking temperature of 450°F, films resisted attack by detergent (1% Tide XK solution at 165°F) for 175 hours. Coulombic yield of the resin system was ca. 46 mg per coulomb. Infrared spectroscopy revealed that the low molecular weight styrene–maleic anhydride copolymer and the amino cross-linking agent migrate to the anode at comparable rates.*

In the electrocoating field many resin systems have achieved technical and commercial success as coatings for electrically conductive substrates. For a compilation of these, together with a bibliography providing an extensive review of electrocoating, the monographs by Chandler should be consulted (1–3). In this laboratory recent work has demonstrated the utility of a partially esterified low molecular weight copolymer of styrene and maleic anhydride (SMA-I) as an electrocoating vehicle which yields films exhibiting excellent corrosion and detergent resistance. Attention was centered on this type of resin because previous work (4) demonstrated these characteristics with solvent-based baking enamels derived from a similar resin. [These copolymers and their partial esters are available from ARCO Chemical Co. under the tradename SMA.]

The synthesis of this type of polymer is outlined in Figure 1 (5) and is based on free radical chain polymerization followed by esterification. As a result of the polymerization process the resins are available as solids. A broad range of resin properties is made possible by varying the ratio of the monomers polymerized as well as the chemical composition of the esterifying alcohol(s) and the extent of esterification. Thus the balance of styrene and maleic anhydride units in conjunction with the esterifying alcohol(s) establishes the carboxyl functionality available for reaction with crosslinking resins.

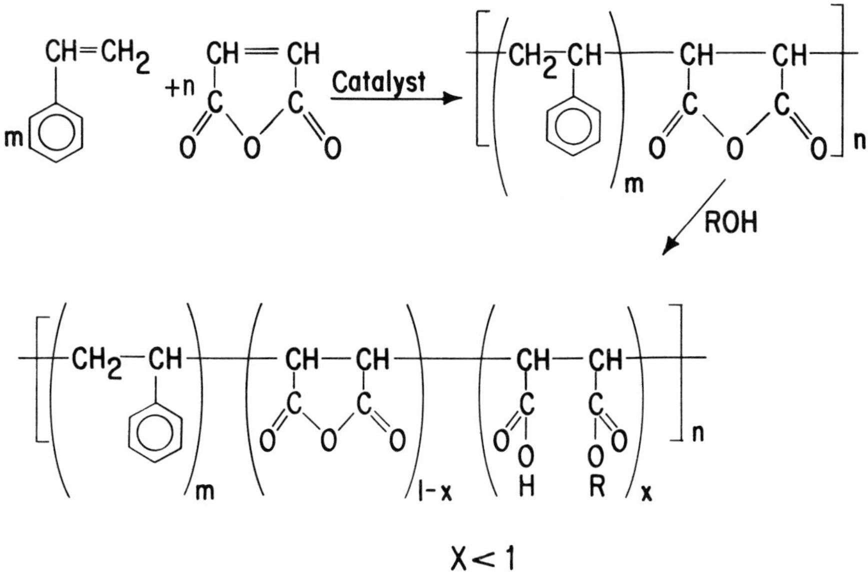

Figure 1. Synthesis of SMA resins

Unlike polyester and alkyd resins, SMA-I type polymers possess a backbone which is stable in the presence of alkali. The hydrolysis of the pendant ester groups in the SMA-I type polymer is not facile (6), particularly at the pH's commonly used in electrocoating. This property is of particular significance for long term electrocoating bath stability. The SMA-I type polymer has a partial acrylate type structure adjacent to carboxyl functionality (7) (Figure 2). The esterifying alcohol plasticizes the polymer and also blocks the adjacent carboxyl group. Steric hindrance of ester groups and any unreacted carboxyl groups would enhance the resistance of the cured film to attack by alkali.

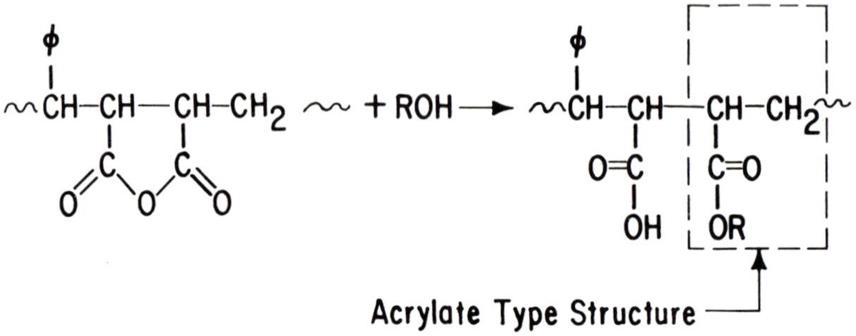

Figure 2. Partial structure of SMA-I

The system described in this paper is SMA-I crosslinked with XM-1123, an amino resin produced by the American Cyanamid Co. All data reported are based on unpigmented films. Pigmentation of the SMA-I system and resultant performance will be the subject of a later paper.

Experimental

Preparation of the Electrocoating Baths and Coated Panels. The electrocoating vehicle was synthesized from a low molecular weight copolymer of styrene and maleic anhydride by reaction with one anhydride equivalent of a long chain alcohol. Examples of esterification of this type of polymer have been disclosed (8). Typical properties of SMA-I are summarized in Table I.

Table I. Properties of SMA-I

Resin Type	Half-Ester of Styrene-Maleic Anhydride Copolymer
Physical form	Solid
Molecular weight	3000–3200
Acid number	100 ± 10 mg KOH/gram
Melting point	80°–100°C
Viscosity at 25°C (70% in Propasol P)	360 poise
Color (50% in methyl isobutyl ketone)	100–125 APHA (ASTM D 1209–62)

The coupling solvent which was used for bath preparation and which has yielded an excellent combination of properties is the monopropyl ether of propylene glycol (Propasol P from the Union Carbide Corp.). Typically, the SMA-I was dissolved in the cosolvent (70–75% nonvolatile), mixed with the crosslinking agent, neutralized (25%) with triethylamine, and dispersed in deionized water with high speed agitation. The water was added gradually until the resin system passed through a phase inversion. The bath solids were adjusted to 10%, and the bath aged for at least 24 hours. Aging may be particularly important for SMA-I electrocoating systems because the carboxyl groups on the polymer chain are shielded by the nearby alcohol moieties. Uncoiling of the polymer chain and attainment of equilibrium of neutralized carboxyl groups can be accomplished by aging. The pH of the freshly prepared bath is 9.0 but drops to 8.1 in 48 hours. Electrocoating was conducted at pH 8.1–8.5 with higher pH's yielding thinner films.

The test panels were cold rolled steel with a Bonderite 37 pretreatment (Hooker Chemical Corp., Parker Division) and were rinsed with deionized water and dried at 250°F for 6 minutes immediately before coating. Electrocoating was conducted at 200 volts for 2 minutes (except where noted otherwise) followed by rinsing with deionized water to remove adhering paint. Drag-out, however, was minimal, and the coated panels could be baked immediately with no adverse effect.

Cured Film Properties. The detergent resistance tests were based on ASTM method D2248–65 using Tide XK as the detergent. Salt fog tests were run according to ASTM method B117-64. After exposure to salt fog the panels were allowed to dry for 30 minutes, and tape (#310 from the 3M Co.) was applied to the scribes. Removal of the tape revealed rust creepage and blistering. If the scribe did not appear clean,

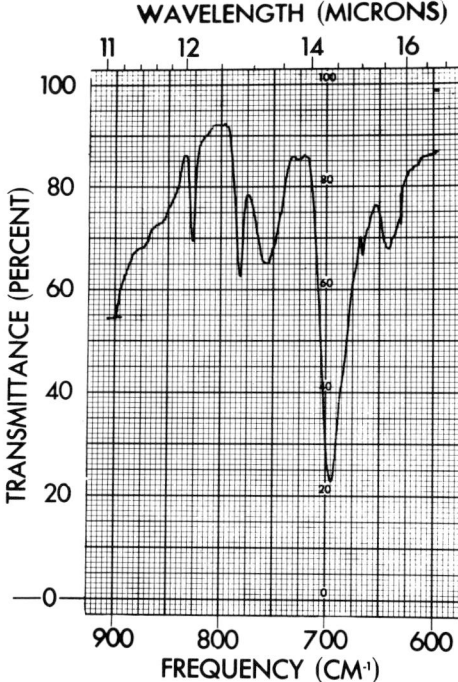

Figure 3. Analysis of films by infrared spectroscopy

the scribe area was scraped with a spatula and retaped. Failure was considered to have occurred when a section of film 1/8 inch wide and 1/4 inch long was removed by the tape. Panels also were considered to have failed when blisters (and/or pinholes) reached the "few" concentration as designated in ASTM method D714-56.

Throwing Power. Throwing power was measured by a published method (9) using a strip of coil coating stock (Bonderite 40) inside a length of 3/4 inch gas conduit. Corrosion protection (10) was determined by submitting the coated strips to salt fog exposure for 240 hours and was designated by the number of inches of corrosion-free area.

Coulombic Yield. Coulombic yield was determined by integrating the amperage decay with time curve using a Speedomax H recorder (Leeds & Northrup Co.).

Deposited Film Composition. The clear films lend themselves to analysis by attenuated total reflectance infrared spectroscopy (11, 12). Several sets of reflectance bands can be used. The ones used in this work were the styrene (ring) band of the SMA-1 at 697 cm^{-1} and the triazine (ring) band of the XM-1123 at 827 cm^{-1}. (Figure 3). For calibration, films of known SMA-I/amino resin compositions were applied to panels with a doctor blade, then cured and analyzed. The compositions were chosen to encompass the range expected from the electrodeposited films. A plot of the ratio of the two net absorbancies vs. amino resin concentration was constructed (Figure 4). Each known was run in

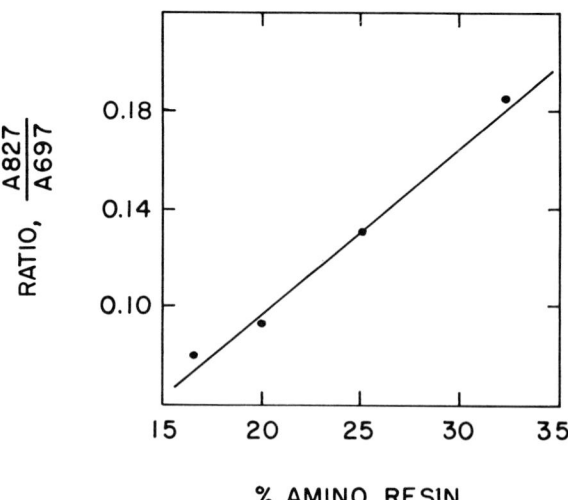

Figure 4. Ratio of the net absorbancies vs. amino resin concentration for two films

duplicate and the results averaged. The thicknesses of the calibration films as well as the curing cycle employed (360°F × 25 minutes) were the same as those for the unknown electrodeposited films.

Results and Discussion

Cured Film Performance. The SMA-I/XM-1123 system exhibits outstanding corrosion resistance as shown by the data in Table II. Maximum

Table II. Effect of Crosslinking Agent Concentration and Curing Cycle on Corrosion Resistance

SMA-I/XM-1123, wt Ratio in Bath	Cure Cycle, °F × min	Salt Fog,[a] hours to Fail	Film Thickness, mil
80/20	450 × 20	480	0.3
80/20	390 × 25	500	0.6
80/20	360 × 25	480	0.5
85/15	360 × 25	450	0.6
75/25	360 × 25	350	0.6
70/30	360 × 25	270	0.6

[a] 5% NaCl fog at 95°F and 100% R.H.

Table III. Effect of Crosslinking Agent Concentration and Curing Cycle on Detergent Resistance

SMA-I/XM-1123, wt Ratio in Bath	Cure Cycle, °F × min	Detergent Resistance,[a] hours with no Effect	Film Thickness, mil
75/25	360 × 25	20	0.8
75/25	390 × 25	28	0.7
75/25	420 × 25	94	0.7
75/25	450 × 20	175	0.6
70/30	450 × 20	140	0.6
80/20	450 × 20	75	0.6

[a] 1% Tide XK solution at 165°F.

corrosion resistance was attained when the SMA-I/XM-1123 weight ratio was ca. 80/20, although excellent results were obtained when this ratio was 85/15. Corrosion resistance decreased as the concentration of the crosslinking agent was increased beyond the 80/20 level.

Optimum detergent resistance was obtained with an SMA-I/amino resin ratio of 75/25. Other concentrations of crosslinker, particularly lower, decreased performance. This is evident from the data in Table

III. Curing temperature also was significant as good detergent resistance was obtained only when the curing cycle approached 450°F × 20 minutes. Data in Table III reflect the exposure during which no significant coating defects were observed. Failure occurred quite rapidly after the specified time. This characteristic is unlike conventional solvent-based coating systems where blistering and rust creepage, once started, progress slowly for the duration of the test. Rust creepage, however, was minimal (less than 1/16 inch) at the scribe, and the films failed by a process of delamination. This phenomenon occurred at the bottom of the test panel and rapidly spread upward. No explanation is offered at this time for this behavior, but it may be caused by some application phenomenon.

Throwing Power. The throwing power tests were run with systems having an SMA-I/amino resin weight ratio of 80/20 and were conducted at different voltages and durations. Baths were aged for 96 hours prior to the test. Throwing powers of 7–8½ inches were obtained (Table IV).

Table IV. Throwing Power

Volts	Time, min	Throwing Power, inches; %	Corrosion Protection, inches	Film Appearance
200	2	7-1/8; 55	6-3/4	Good
200	4	8-3/8; 64	8	Good
300	2	7-3/4; 60	7-1/8	Few Pinholes

Table V. Migration Rate of SMA-I and XM-1123

Weight Ratio SMA-I/XM-1123		% SMA-I in Film / % SMA-I in Bath
Bath Composition	Baked Film Composition	
85.0/15.0	83.9/16.1	0.93
75.0/25.0	73.1/26.9	0.92
70.0/30.0	68.5/31.5	0.95

The percent throwing power was calculated from the immersed depth of the pipe, which was 13 inches. After curing at 360°F for 25 minutes, the strips were submitted to 5% salt fog exposure for 240 hours. Corrosion protection was determined by measuring the length of strip which was not corroded. In all cases corrosion protection was excellent, but some pinhole corrosion was evident in the strip coated at 300 volts. Some rupturing may have occurred at this higher potential. This system exhibited exceptional corrosion protection even in comparatively thin films of 0.1–0.3 mil.

Coulombic Yield. After an aging period of 96 hours coulombic yields were determined for baths containing SMA-I/amino resin ratios of 80/20 and 75/25. Four determinations yielded only marginal differences. The bath containing the greater amount of XM-1123 gave a slightly higher value: 48.6 mg/coulomb *vs.* 46.0 mg/coulomb.

Migration Rates of SMA-I Compared with Crosslinking Agent. It is apparent from Table V that the crosslinking agent exhibited a migration tendency approximately equal to that of the SMA-I. This tendency did not appear to be concentration dependent throughout the range of ratios used.

Acknowledgment

The permission of ARCO Chemical Co. to publish this paper is gratefully acknowledged. Thanks are also due to John Moirano for the infrared analyses.

Literature Cited

1. Chandler, R. H., "Electrophoretic Paint Deposition," R. H. Chandler, Ltd., London, 1970.
2. Chandler, R. H., "Advances in Electrophoretic Painting, 1967–68," R. H. Chandler, Ltd., London, 1969.
3. Chandler, R. H., "Advances in Electrophoretic Painting, 1969," R. H. Chandler, Ltd., London, 1969.
4. Marion, D. L. *et al.*, U.S. Patent **3,528,935** (1970).
5. Gower, B. G., *Detergent Age* (April 1968) 22.
6. Marion, D. L., Motier, J. F., unpublished results.
7. Solomon, D. H., "The Chemistry of Organic Film Formers," p. 264, Wiley, New York, 1967.
8. Muskat, I. E., U.S. Patent **3,342,787** (1967).
9. Brewer, G. E. F., Horsch, M. E., Madarasz, M. F., *J. Paint Technol.* (1966) **38**, 452.
10. Brewer, G. E. F., Strasberg, G. G., Horsch, M. E., *J. Paint Technol.* (1967) **39**, 551.
11. Anderson, D. G., Tessari, D. J., *J. Paint Technol.* (1970) **42**, 119.
12. Koral, J. N., Blank, W. J., Falzone, J. P., *J. Paint Technol.* (1968) **40**, 156.

RECEIVED May 28, 1971.

6

Preparation of Electrodeposition Resins: Lactone Formation During Maleinization

LE ROY S. FORNEY[1] and THOMAS J. SHEERIN

Central Research Laboratories, Mobil Chemical Co., Edison, N. J. 08817

> *The reaction of more than 1 mole of maleic anhydride with unsaturated fatty acids and esters at elevated temperatures (maleinization) was investigated. The previously observed phenomenon of an apparent loss in acid content during maleinization with concurrent elimination of a nonequivalent amount of carbon dioxide was interpreted as largely resulting from condensation of two succinic anhydride moieties, leading to the formation of spirodilactone groupings. The maleinization reaction sequence was studied by IR using model systems and then correlated with carbon dioxide generation during maleinization of commercial fatty acid mixtures.*

The reaction between maleic anhydride and unsaturated systems is well known, and the process, when applied to fatty acids or their derivatives, may be referred to by the term "maleinization." The reactions which occur during maleinization fall into two major categories. With conjugated systems, such as eleostearic acids (1), licanic acids (2, 3), or 8,10- (4), 9,11- (5), and 10,12- (4) octadecadienoic acids and their derivatives, a Diels-Alder addition occurs at 60°–100°C. In this process a 1,4- addition provides a cyclohexene derivative with loss of one of the conjugated double bonds. With systems containing isolated double bonds such as oleic acid and its derivatives (6), on the other hand, temperatures around 200°C are required. The products conform predominantly to the ene reaction (7), involving addition of maleic anhydride to an olefinic substrate possessing an allylic hydrogen (unconjugated fatty acid or derivative) with an allylic shift of the substrate double bond.

[1] Present address: Mobil Oil Corp., 150 East 42nd St., New York, N. Y. 10017

Diels-Alder Reaction:

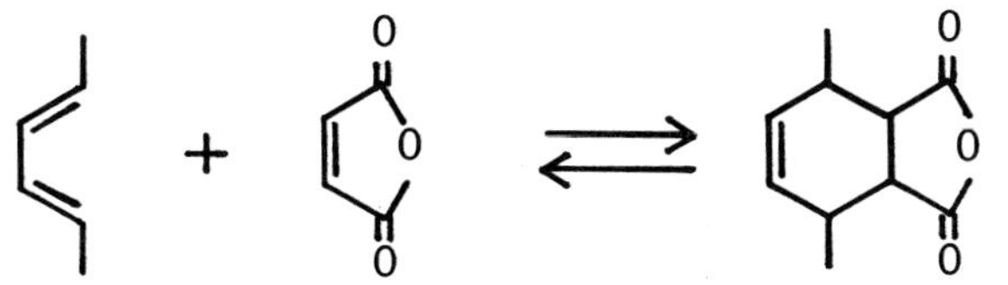

Ene Reaction:

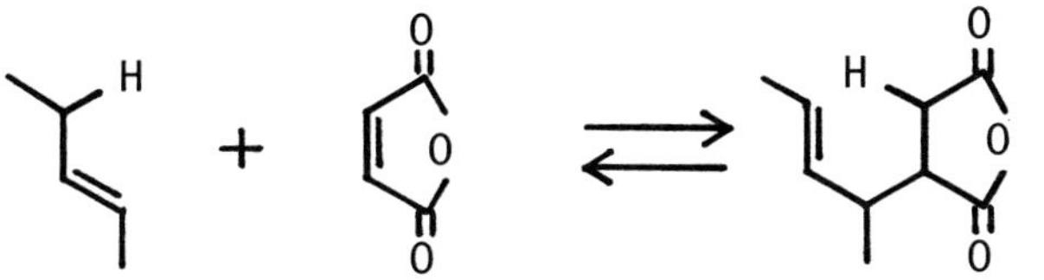

A report that at least some maleic anhydride addition occurs allylic to unconjugated saturation but without shifting the double bond (8) has been interpreted as an indication of a free radical mechanism leading to products.

The maleinization reaction is important in preparing electrodeposition resins, where introduction of free carboxylic functionality is essential for transport phenomena. The ease and economics of maleinization guarantee that it is, and will continue to be, used widely to modify resins for electrodeposition. Its use in epoxy ester systems is but one example of its wide utility in electrodeposition resins. A commercial system of this type involves reaction of a bisphenol-A epichlorohydrin copolymer with linseed fatty acids and functionalization of the product by maleinization (DX-15 resin, Shell Chemical Co.). Since electrical properties as well as final resin properties are likely to be determined in part by the method of introducing the transport functionality, a clear understanding of the maleinization reaction is desirable.

While the major maleinization processes described earlier are well understood for the addition of 1 mole of maleic anhydride to an unsaturated fatty acid system, the situation is much less clear when the reaction involves several moles of maleic anhydride. A study of such systems has shown that both succinyl and cyclohexane moieties are present in product mixtures from the reaction of maleic anhydride with multi unsaturated fatty acids. These results clearly indicate concurrent Diels-Alder and ene reactions leading to complex product distributions (9). Nonetheless, these reactions alone cannot account for the apparent loss of acidity which occurs during maleinization of multi unsaturated fatty acids or derivatives. In view of the importance of maleinization as a method for introducing acid functionality into electrodeposition resins, we investigated this phenomenon of acidity loss in some detail.

Experimental

Linoleic acid was prepared from safflower oil (10) and purified through the tetrabromide (11). The methyl ester was prepared by treating the acid with 10% BF_3CH_3OH. Acid and ester were free of IR (infrared) absorption at 968 cm^{-1}, characteristic of an isolated trans double bond (11, 12), and the ester gave only one peak on gas chromatography.

Maleinizations were conducted by heating the acid (7.10 mmoles) or ester (5.76 mmoles) with three equivalents of maleic anhydride in an oil bath at 200°C for 2 hours and 240°C for an additional hour, under nitrogen, with occasional shaking. IR spectra of these materials after cooling indicated the presence of excess anhydride (absorption at 3100 and 697 cm^{-1}), which was removed by heating the samples to 100°C/0.15 mm for 15 minutes. Linoleic acid weight increase: Calc. for dianhydride, 69.9%; for spirodilactone, 52.8%; Found, 64.3%. Methyl linoleate weight increase: Calc. for dianhydride, 66.6%; for spirodilactone, 50.3%; Found, 57.3%.

Decarboxylation of maleinized methyl linoleate was done by the sodium hydride treatment described in the literature (13).

Half-esters of maleinized methyl linoleate and linoleic acid were formed by refluxing the maleinized ester (0.181 gram) and acid (0.185 gram) with 5 ml absolute methanol, periodically removing samples for IR analysis.

Commercial fatty acids were maleinized by heating a mixture of the fatty acid and maleic anhydride (1/1 mole ratio) to 200°–230°C for 4–5 hours, as shown in Table I. The apparatus was swept with a stream of CO_2-free nitrogen, and the CO_2 evolved was scrubbed free of water and carboxylic acids, then absorbed in a gas train and weighed. The apparatus is shown schematically in Figure 1. In all cases, the reactions

Table I. Theoretical and Actual Product Acid and Saponification

| | | | | \multicolumn{8}{c}{Acid Number} |
| Substrate | \multicolumn{3}{c}{Maleinization} | \multicolumn{2}{c}{Measured} | \multicolumn{6}{c}{Calculated Values} |

	Time, hr	Temp, °C	CO_2 Loss, wt %	Aq	Alc	SDL[a] Aq	Alc	Keto-acid[b] Aq	Alc	Decarbox.[c] Aq	Alc
LOFA	4.3	200–230	1.51	344	263	374	263	413	283	432	282
SOFA	5.4	200–220	1.97	314	248	351	251	402	277	429	277
DCOFA	5.0	200–220	1.22	313	260	387	268	418	284	404	284

[a] SDL = spirodilactone. Formation of spirodilactone requires loss of four acid groups toward aqueous titration and loss of two acid groups toward alcoholic titration, per mole of CO_2 generated.

[b] The keto-acid is assumed to decarboxylate. Two acid groups are lost toward aqueous titration, and one group is lost to alcoholic titration, per mole of CO_2 generated.

[c] Direct decarboxylation; one acid group is lost per mole of CO_2 generated.

were continued until the free maleic anhydride level was <1%, as estimated by color formation with N,N-dimethylaniline and comparison with standards. A typical reaction is detailed below for 741 grams (2.646 moles) SOFA with 259 grams (2.643 moles) maleic anhydride (Table II).

Acid Number (mg KOH/gram sample). AQUEOUS. Tetrahydrofuran/water (50/20) diluent. Aqueous KOH titrant, Phenolphthalein indicator. Titrates MA as two equivalents of acid.

ALCOHOLIC. Toluene/methanol (50/50) diluent. KOH in absolute methanol titrant, Phenolphthalein indicator. Titrates MA as one equivalent of acid.

SAPONIFICATION NUMBER. The usual method was modified to accommodate the greater reactivity of the esters formed. Titrate as usual to an aqueous AN endpoint, then heat to just below reflux with nitrogen blanket and titrate till endpoint holds for one-half hour.

Calculations. THEORETICAL AN:

$$\text{Theoretical AN} = \frac{[\text{Equiv. charged} - \text{moles } CO_2(X)] (56.1)}{(\text{Charge weight} - \text{weight } CO_2) 10^{-3}}$$

where X varies as follows:

	Aqueous	Alcoholic
Spirodilactones	4	2
Decarboxylated keto-acid	2	1
Decarboxylation-direct	1	1

THEORETICAL SAPONIFICATION NUMBER. Assuming that formation of spirodilactone is responsible for the drop in acidity from 445 at the start to some final value:

$$\text{Theoretical Sap. No.} = (445 - ANaq)/2 + ANaq$$

Numbers for Fatty Acid/Maleic Anhydride Reactions, 1/1 Molar

Saponification Number			
Measured	Calculated Values		
	SDL^d	Keto-Acid[e]	Decarbox.[e]
400	395	344	344
378	380	314	314
383	379	313	313

[d] Spirodilactone formation would result in the loss of two acid groups, which is only half the loss experienced by the aqueous AN. Calculated saponification number was based on the initial and final aqueous AN (see Experimental).

[e] No saponfiable ester groups, so the saponification number should equal the aqueous AN.

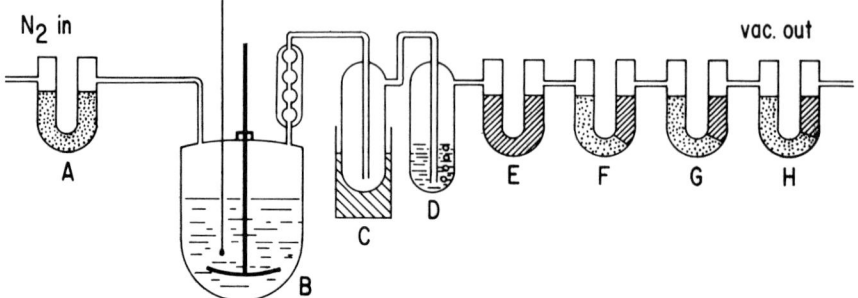

Figure 1. Apparatus for trapping CO_2 during maleinization
A = Ascarite tube to remove traces of CO_2 from N_2 stream
B = reactor equipped with thermometer, stirrer, and H_2O condenser
C = dry ice trap
D = bubble meter (nujol)
E = anhydrone ($MgClO_4$) to remove H_2O and acids from gas stream
F, G, H = 2/3 Ascarite, 1/3 anhydrone to trap evolved CO_2. All weight increases were found in tube F.

Results and Discussion

Maleinization Products. In the reaction of maleic anhydride with linoleic acid or its ester, two moles of anhydride react per mole of substrate (14, 15, 16). Amounts of CO_2 generated are insufficient to explain this acidity loss as simple decarboxylation. Consequently, reactions between an anhydride and a reactive methylene group to form keto-acids which subsequently decarboxylate have been proposed (15) as shown below.

However, a review of the ene reaction (17) cited an alternative course for further reaction of ene products with loss of CO_2 where two or more anhydride groups are in close proximity. Thus, maleinization of 1,4-pentadiene may be expected to yield an ene product with a single anhydride grouping. However, this product cannot be isolated since the allylic shift of the double bond provides a conjugated system which immediately adds another maleic anhydride unit by a facile Diels-Alder reaction (18). This product may then lose CO_2 to form a spirodilactone compound, as shown below (13). This loss of CO_2 may occur spontaneously or under the influence of basic catalysts. In a formally analogous process, the pyrolysis of succinic anhydride provides the spirodilactone of γ-ketopimelic acid (19).

The molecular reorganization which results in the loss of CO_2 in the reactions noted above has been compared with such base-catalyzed processes as the Claisen or Perkin reactions (13). Accordingly, reaction would involve formation of a carbanion adjacent to an anhydride car-

Table II. Maleinization of SOFA

Time, hr	Temp., °C	Acid Number				CO_2 Yield, grams	% Free MA
		Aqueous		Alcoholic			
		Found	Calc.	Found	Calc.		
0	25	443	445	296	297	0	100
0.42	160	405		299			
0.75	210						
0.58	220						
2.83	220	368	392	266	271	11.26	2
4.50	220	330	361	253	256	17.74	1
5.42	220	314	351	248	251	19.72	<1

bonyl group, which undergoes two rearrangement steps to produce a tertiary carboxyl group. The tertiary carboxyl group then undergoes characteristic decarboxylation under the reaction conditions. This sequence is shown below for the diadduct of 1,4-pentadiene.

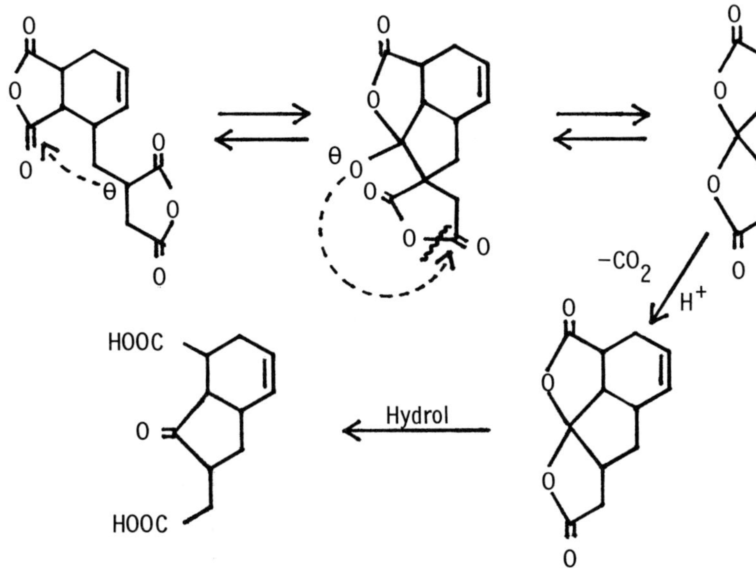

If spirodilactones are formed by a similar process on decarboxylation of fatty acid systems during maleinization, the characteristic spirodilactone IR absorption at 1795 cm^{-1} should be discernable. Therefore we treated methyl linoleate with maleic anhydride under nitrogen for three hours at 200°–240°C. The IR spectrum of the product exhibits strong anhydride absorption at 1780 and 1860 cm^{-1}. It does not show a band attributable to spirodilactone at 1795 cm^{-1} although this peak could easily be masked by the intense anhydride absorption at 1780 cm^{-1}. Therefore, we treated the product with a strong base, under conditions which effect the decarboxylation (13). After workup, the anhydride absorptions at 1780 and 1860 cm^{-1} were completely absent, and a new sharp peak appeared at 1795 cm^{-1}. Thus, maleinized methyl linoleate appears susceptible to spirodilactone formation, at least under basic conditions.

It remained to be determined, however, whether spirodilactones may result directly from the maleinization process without intervention of a base-treatment step. Therefore, we sought a process which would remove anhydride functionality without appreciably destroying spirodilactone functionality. The facile uncatalyzed reaction of anhydrides with alcohols to produce half-esters provided a suitable technique. Figures 2 and 3 present successive IR spectra taken during the uncatalyzed reaction of maleinized linoleic acid and methyl linoleate with refluxing methanol. Both show a rapid decrease in the anhydride peak at 1780 cm^{-1}. After one hour this peak has decreased and broadened significantly, and by two hours its maximum has shifted to 1795 cm^{-1} (Figure 2), or two peaks at 1780 and 1795 cm^{-1} are clearly visible (Figure 3). In both cases, the smaller anhydride peak at 1865 cm^{-1} diminishes steadily. After two hours, peaks attributable to both anhydride and spirodilactone slowly disappear, finally leaving only COOH and COOR absorption after 50 hours. Methyl esterification of the reactants or products catalyzed by mineral acids or BF$_3$ leads rapidly to a single peak at 1740 cm^{-1}. However, since the ketone expected from esterification of the spirodilactone and the ester groups in the system both absorb in this region, no further information about the reaction can be gained from these data.

Having shown that spirodilactones may result from maleinization of linoleic acid and methyl ester, it seemed valuable to analyze the maleinization products derived from commercially available fatty acids. The generation of CO$_2$ from linseed acids (LOFA), soya acids (SOFA), and dehydrated castor acids (DCOFA) was monitored and correlated with

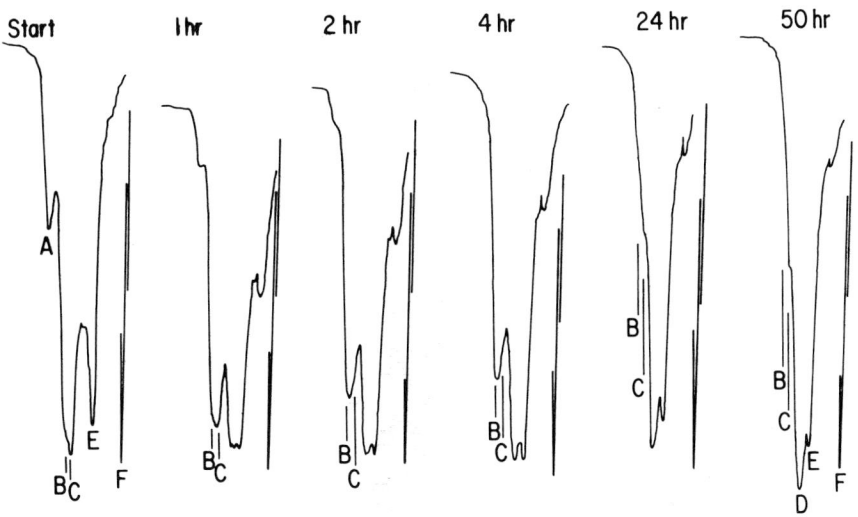

Figure 2. IR study of maleinized linoleic acid refluxed in methanol

A = 1860 cm^{-1}: anhydride absorption
B = 1795: spirodilactone
C = 1780: anhydride
D = 1740: ester
E = 1710: carboxylic acid
F = 1601: polystyrene marker for spectral calibration

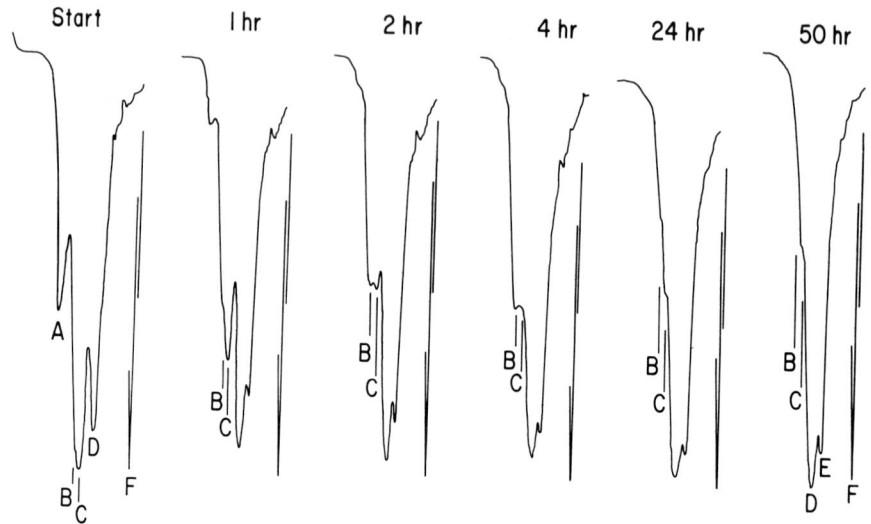

Figure 3. IR study of maleinized methyl linoleate refluxed in methanol

$A = 1860\ cm^{-1}$: anhydride absorption
$B = 1795$: spirodilactone
$C = 1780$: anhydride
$D = 1740$: ester
$E = 1710$: carboxylic acid
$F = 1601$: polystyrene marker for spectral calibration

the drop in acid and saponification numbers. These acids were chosen because of their availability, differing amounts and types of unsaturation, and their importance in preparation of electrodeposition resins.

The three major pathways proposed for decreasing acidity upon maleinization were compared, using these fatty acid systems. The pathways include direct decarboxylation (14), keto-acid formation followed by decarboxylation, and spirodilactone formation. The results in Table I are consistent with the occurrence of spirodilactone formation as the source of the CO_2 which is produced in the reactions.

For LOFA and SOFA the excellent agreement between the calculated and experimental acid numbers supports the theory of spirodilactone formation. For the partially conjugated system DCOFA, the agreement is not as good, but it is still more consistent with spirodilactone formation than with the other proposed products. The aqueous acid numbers are somewhat lower than predicted from the amount of CO_2 collected, but again the best agreement is with spirodilactone formation. In addition, the ester present is very reactive, as indicated by an endpoint that disappears much faster than is common in acid titrations in the presence of ester—a fact that would be consistent with the presence of spirodilactones which are likely to be unusually susceptible to saponification or hydrolysis.

Summary

We confirmed the observation of other workers that maleinization of unsaturated fatty acids and esters is accompanied by large losses in acidity and generation of CO_2. In addition we found that an unusually reactive ester is formed. Several reactions have been proposed in the past to explain these phenomena, but none fits the experimental data very well.

In maleinization of simpler diene substrates, spirodilactones are formed. To determine if these were also formed in fatty acid systems, we investigated the behavior of maleinized products by IR and correlated the drop in acidity with CO_2 generation and saponification number. The results are in accord with the postulated dilactone formation and account for a significant portion of the maleinization products obtained in this study.

The presence of spirodilactone may account for the fading endpoints which often characterize basic titration of maleinization products. It would also be expected that in water-based systems, such as those used for electrodeposition, spirodilactones would undergo hydrolysis. This could adversely affect such acidity related phenomena as bath stability, transport, and cured-film properties.

Literature Cited

1. Morrell, R. S., Samuels, H., *J. Chem. Soc.* (1932) 2251.
2. Morrell, R. S., Davis, W. R., *J. Chem. Soc.* (1936) 1481.
3. Morrell, R. S., Davis, W. R., *Trans. Faraday Soc.* (1936) **32**, 209.
4. Von Mikusch, J. D., *J. Amer. Oil Chemists Soc.* (1952) **29**, 114.
5. Boeseken, J., Hoevers, R., *Rec. Trav. Chim.* (1930) **49**, 1165.
6. Ross, J., Gebhart, A. I., Gerecht, J. F., *J. Amer. Chem. Soc.* (1946) **68**, 1373.
7. Alder, K., Pascher, F., Schmitz, A., *Ber.* (1943) **76**, 27.
8. Bickford, W. C., Fisher, G. S., Kyame, L., Swift, C. E., *J. Amer. Oil Chemists Soc.* (1948) **25**, 254.
9. Rheineck, A. E., Khoe, T. H., *Fette, Seifen, Anstrichmittel* (1969) **71**, 644.
10. McCutcheon, J. W., *Org. Syn.*, Coll. **III**, 526, note 7.
11. Jackson, J. E., Lundberg, W. O., *J. Amer. Oil Chemists Soc.* (1963) **40**, 276.
12. Ahlers, N. H. E., Brett, R. A., McTaggart, N. G., *J. Appl. Chem. (London)* (1953) **3**, 433.
13. Hill, R. K., Barger, H. J., Jr., *J. Org. Chem.* (1965) **30**, 2558.
14. Bickford, W. G., Krauczumas, P., Wheeler, D. H., *Oil Soap* (1942) **19**, 23.
15. Plimmer, H., *J. Oil Colour Chemists Assoc.* (1949) **32**, 99.
16. Brett, R. A., *J. Oil Colour Chemists Assoc.* (1964) **47**, 767.
17. Hoffmann, H. M. R., *Angew. Chem., Internat. Edit.* (1969) **8**, 556.
18. Alder, K., Munz, F., *Ann.* (1959) **565**, 126.
19. Moldarskii, B. L., Blinova, M. V., *Neftekhimiya* (1965) **5**, 108; *Chem. Abstr.* (1965) **62**, 16043b.

RECEIVED May 28, 1971.

7

Pigmentation of Electrocoatings

DAVID S. YOUNG and ARTHUR T. GRONET

MPM Division, Pfizer Inc., 640 North 13th St., Easton, Pa. 18042

> *While it is generally agreed that the vehicle influences the performance of a pigmented electrocoat system more than the pigment, certain pigments do have a significant influence on many parameters of electrocoating. Particle size of the pigment, assuming good dispersion, has little effect on performance. A pigment with high specific resistance will usually transfer more efficiently and give better bath stability. Increasing the pigment-to-binder (P/B) ratio will generally increase the specific resistance and throwing power of the bath but will have little effect on coulomb efficiency or rupture voltage whereas the film becomes harder and the gloss decreases especially at P/B's above 1/1. Pigments suitable for electrocoat should have good alkali resistance, low solubility in water, easy dispersibility, minimum settling, controlled particle size, and a balance between their specific resistance and their specific conductivity in aqueous suspensions that allows good mobility.*

While the operation of electrodeposition systems may be relatively simple, the mechanism of deposition is complex, involving electrolysis, electrophoresis, electroosmosis, oxidation of metal, coalescence of colloidal particles, and gas formation. The binder of an electrodeposition coating can be a modified oil, alkyd, acrylic, or an epoxyester resin. The polymers used for anodic deposition have free acid or acid anhydride groups attached to the polymer chain. Although these binders may be used alone and clear films deposited, this paper deals strictly with pigmented systems.

An electrodeposition primer and/or topcoat may be visualized as a suspension of pigment particles and polymer-coated pigment particles with the ionized carboxyl groups covering the surface so that the particle carries a negative electrical charge. Since like electrical charges repel each other, these negative charges are responsible for keeping the particles dispersed in the water. These charges also enable the particles to migrate toward the positively charged anode during the electrodeposition process when a direct current of 50–400 volts is applied. The film is formed by coagulation at the anode of these negatively charged macromolecules of the binder. At the same time this film is made practically anhydrous by electroosmosis.

The advantages of the electrocoating process over conventional methods of paint application are: (1) uniform pinhole-free coatings, (2) deposition of greater film thickness on hard-to-reach recessed areas and areas of broken contour, thus increasing corrosion resistance, (3) automated operation, (4) elimination of fire hazards and air pollution problems, (5) low operating costs, and (6) fast throughput. The main disadvantages of the electrocoating process are: (1) high cost of installation, (2) complex quality control measures necessary to ensure good stability of the operating tank, and (3) the fact that the system is presently limited to one-coat applications and dark or pastel colors.

Scope

The electrodeposition of primers and one-coat paint systems as industrial production finishes has been proved a practical and beneficial commercial method of coating production line products. Major areas of present interest are (1) automotive primers, (2) appliance primers, and (3) general purpose one-coat systems.

As of September 1969 more than 400 electrocoat installations were operating in various industrial plants around the world (1). These installations were coating about 18 million square feet of metal surface per day. These 400 production tanks were split up on the five continents as follows:

	Number of Installations (1)	Output, sq ft/day (1) (Estimated)
Africa	5	36,000
America	80	2,900,000
Asia	35	1,450,000
Australia	3	36,000
Europe	282	14,400,000
	405	18,822,000

Information available on the 282 European tanks shows that the following types of articles were electrocoated:

	Number of Installations	In., %	Output sq ft/day	In., %
Automotive industry	71	26	11,000,000	76.0
Automotive accessories	46	16	544,000	4.2
Steel furniture	26	9	300,000	2.2
Electro-industry	25	9	240,000	1.6
Others	114	40	2,400,000	16.0
	282	100	14,484,000	100.0

The number of electrocoat tanks installed and the resulting paint capacity in electrocoat operations in North America increased tremendously from 1966 to 1971. The total capacity of electrocoat tanks in North America was only 81,000 gallons in 1966 whereas in 1971 it was approximately 2 million gallons. A breakdown of tanks, tank capacity, and paint usage in 1971 for North America is shown in Table I (2).

Table I. Electrocoat Tanks and Usage in North America

	Tanks	Range	Capacity, 1000 gal Average	Capacity, 1000 gal Total	Paint, million $
Autos and trucks	32	8–76	38	1180	9
Auto parts	16	1.5–24	6	90	2
Electrical	35	0.2–18	5.5	210	4
Appliances	26	3–22	9	220	4
Aluminum extrusions	6	0.3–30	8	40	1
Miscellaneous	27	2.8–11	4.5	250	5
Total	142			1990	25

The latest figure for 1972 shows at least 160 tanks in operation in North America, using about 2.2 million gallons with a paint value of about 28 million dollars (2). It is estimated that approximately 20% of the passenger cars are primed by electrocoat in the United States, 40% in Europe, and over 90% in Japan. Therefore, at present, the automotive industry certainly is the major user of electrodeposited paints.

Historical Background

A great deal of information pertinent to the problem of electrophoretic paint deposition can be obtained from the studies reported and the literature on the electrodeposition of latex and various resins and related compositions. One of the most significant of the literature references (3) reports that two processes for depositing rubber electrically from a latex were developed in the period 1925–1927. One was developed by Sheppard and Eberlin and the other by P. Klein, and a number of patents were issued. Both processes were based on the fact that the dispersed particles of rubber in the latex carry a negative charge and in an electric field migrate to and are deposited on the anode where they coagulate. This anaphoresis principle was known to Knox in 1907 and Cockerill in 1908, and they attempted to coagulate rubber from the latex by this method.

One of the first industrial applications of electrophoreses was in the dewatering of ceramic clay. This was developed by Count Schwerin in Germany in the early 1900's. With an electric input of 80 kwh/ton of dry clay is was possible to dewater a slurry from 35% solids suspension to approximately 65% solids plastic clay. The cell potential was normally about 100 volts. Although a higher voltage gave a dryer product, more energy was lost by the electrolysis of water and heating of the cell. Both electrophoretic movement of the clay particles to the anode and the electroosmotic movement of the water to the cathode take place in the cell.

The deposition of plastic, electrophoretically, from nonaqueous solutions has been studied by Feinlieb (*4*). In this study the resin or plastic was taken into solution in an appropriate organic solvent, and a non-solvent diluent was added to cause precipitation as a disperse phase. Under such conditions the dispersion would be stable for at least the time required to run the tests. Feinlieb studied a vinyl chloride–vinyl acetate copolymer which was electrophoretically deposited from a solution of butyl acetate to which 95% ethyl alcohol was added as the precipitant. A good adherent film was obtained. In a later and more detailed study, Fink and Feinlieb investigated the electrodeposition of a number of synthetic resin lattices (aqueous dispersions) obtained as commercial products from the manufacturers. The lattices tested included:

(1) Copolymer of vinylidene chloride and acrylonitrile
(2) A styrene–butadiene copolymer
(3) An unplasticized poly(vinyl chloride)
(4) A plasticized poly(vinyl chloride)
(5) Poly(vinyl acetate)

The "negative" lattices—*i.e.*, all except the poly(vinyl acetate)—gave good deposits on the anode under satisfactory conditions: voltages on the order of 1 volt and current densities on the order of 25 ma/sq inch. A cell was used in which the cathode was separated by a diaphragm from the bath to prevent hydrogen gas bubbles from being carried to the anode and being deposited along with the film.

Deposits of cellulose were obtained by Mantell and Cozzarelli from 1% solution of cellulose in the sodium zincate. The anode deposit was obtained with cell voltages from 1.1 to 1.28 volts and a current density of *ca*. 1–7 ma/sq inch.

Deposits of metal powders and inorganic refractory carbides and oxides have been obtained electrophoretically from nonaqueous suspension mediums. In ane study by Mosley and Wallace (5) the liquids used were isopropyl alcohol and nitromethane. The adhesion of the anode deposit was improved by adding $0.15M$ ammonium hydroxide (using isopropyl alcohol to dilute the concentrated aqueous ammonia solution).

In a somewhat different study by Pearlstein, Wick, and Gallaccio (6), an attempt was made to deposit electrophoretically finely divided metal powders of such metals as aluminum, titanium, zirconium, and tungsten to obtain metal coatings in those cases where ordinary electroplating methods cannot be used. It was found that aluminum electrophoretic deposits could be obtained at voltages ranging from 20 to 320 when using as suspending medium aliphatic alcohols having more than three carbon atoms. Ethers, esters, and ketones were unsatisfactory. In the case of aluminum the suspending medium used determined whether metal was deposited on the anode or cathode, and the addition of a small amount of butylamine greatly improved the adhesion of the deposit.

The deposition of fine mica flakes was studied by Hirayama and (7) Berg. A silicone binder was used, and both silicone and mica plated out together at a pH of 8 to 9.5. This system is sensitive to pH.

Technical Aspects of Pigmented Electrocoating Vehicles

The movement of both pigment and vehicle particles in an electrocoat system is associated with the "electrokinetic phenomenon." This phenomenon includes all processes in which electrically charged particles are acted upon by an external electrical field which results in the charged particles moving through, or relative to, a fluid medium. If the particle is not free to move, the system becomes mechanically strained. The development of knowledge in this area closely parallels the study of colloids and is said to have been discovered by Reuse in 1808. From about 1910 to 1935, many theoretical studies were made in this area, and a number were the fundamental basis for much of the modern work in electrophoresis.

Emulsions, colloids, and sols consist of electrically charged particles of a range of sizes dispersed in a medium whose dielectric properties prevent the charges from being dissipated. The properties of the medium in which the charged particles are dispersed greatly influence the properties of the system. If the medium has a high dielectric constant, as does water, the movement of very small charges (ions for example) takes place easily at potentials of less than a couple of volts. However, if the medium is a nonpolar hydrocarbon, field potentials of several hundred to a thousand volts may be necessary to show particle movement. If the viscosity is low, as in water, the rates of motion of charged particles will be much greater than in a viscous medium (for the same electrical field).

The movement of a charged particle suspended in a medium which is induced by an external electric field is actually movement relative to the medium. If the charged particle is fixed, as in porous membranes, the liquid will move through the membrane. Conversely, if liquid is caused to flow through a porous membrane or capillaries, the membrane or capillaries acquire electrical charges. Even particles settling out of a suspension will acquire an electrical charge from movement through the liquid medium.

The particles of a colloid remain in suspension by reason of the electrical charges on each particle. Some colloids have particles whose charges are either positive or negative. As a result, the particles of one colloid may migrate to the anode, and of another colloid to the cathode. The mixing of oppositely charged colloids usually results in precipitation although different systems vary greatly in sensitivity. The addition of an electrolyte to a colloid will often cause the colloid to precipitate by increasing the conductivity of the medium and permitting the colloid charges to be neutralized or dissipated.

Many of the properties of the charged particles in dispersions are related to the electrical "double layer." Each negatively charged particle, for example, will be surrounded by a sheath of positive charges, resulting from the repulsion of electrons from its neighborhood in the enclosing medium. Such particles behave as a charged capacitor, and this aspect of porous diaphragms, electrodes, etc., has been studied extensively.

The weight of matter transported for each coulomb of electricity passed through a colloid system follows Faraday's Law although it is not as simple to determine in this case as it is in the case of ions in electrolysis.

When it is considered that in a colloid the particles are present in a range of sizes, with the smallest many times the size of ions, and that each particle may have a range of electrical charges on the order of hundreds of thousands of electrons, it is to be expected that average values of weight and charge would need to be determined. This is difficult since these factors would vary with age, method of preparation, and other factors. In electrolysis, each ion of the species can be considered as having a weight and a charge identical to all the other ions of that species, but with the colloids, statistical values must be used.

The basic mathematical relation for electrokinetic phenomena is credited to Helmholtz:

$$u = \frac{e\,Z\,E}{4\,\pi\,n}$$

u = migration velocity of charged particle (pigment or vehicle)
Z = electrokinetic or zeta potential
E = applied voltage
   dielectric constant of the liquid phase
n = viscosity of the medium
4π = a factor related to shape and size.

For ions or very small spherical particles, the value approximates 6. For particles of larger size, the smaller factor is said to be caused by the distorting effect of an insulating particle in the electric field.

The equation shows that the rate of pigment or vehicle particle movement (electrophoresis) or flow of liquid phase (electroosmosis) is proportional to the applied voltage, E. The zeta potential may be defined as the voltage difference between the particle and the diffuse potential value of the double layer surrounding it.

Influence of Pigment on Deposition Characteristics

Particle Size. Studies of synthetic and natural iron oxides and extender pigments such as talc, barytes, and calcium carbonate have shown that the particle size of these pigments (which ranged from 0.5 to 8 microns in diameter) had little or no effect on their transfer properties in electrocoat primer systems. However, pigments which have a tendency to flocculate or agglomerate will have lower mobility rates. Generally, the higher the resistance properties of the electrocoat vehicle, the better will be the transfer rates of the pigment.

Effect of pH. Since all of the electrocoat systems in general production use operate in a pH range of 7.5 to 9.5, a study of synthetic and natural iron oxides in water slurries at pH levels of 7.0 to 10.0 showed that as pH increases, the pigment mobility decreases slightly, and at pH levels over 10, some oxides appear to become neutral in charge or even reverse charge. Generally, electrocoat primers increase in conductivity, have lower throwing power and film thickness, increased gassing, and draw greater current as the pH of the bath is increased.

Settling Characteristics. Because of the low viscosities at which electrocoat tanks or baths are operated, settling out of the pigment and/or vehicle is a serious problem. All large commercial tanks use either agitators or elaborate pumping equipment to maintain a uniform tank. As in conventional paint systems, pigments which have a high specific gravity and which are difficult to grind will cause the most settling problems in electrocoat tanks. However the manner in which the dispersed pigmented paste grind is let back and the resulting stability of the reduced electrocoat paint system in regard to preventing flocculation play an important part in the settling characteristics of an electrocoat system. Generally about 1% of a beneficiated natural hectorite, bentonite, or synthetic clay may be added to an electrocoat primer system to help the suspension properties.

Effect of Pigment/Binder Ratio on Electrical Resistance. Usually the specific resistance of an electrocoat primer system will increase when the pigment-to-binder ratio is increased. This increase will usually be related to the specific resistance and conductivity of the prime and/or extender pigments used to increase the pigment level.

COULOMB EFFICIENCY. Increasing the PVC content of an iron oxide electrocoat primer will generally decrease the conductivity slightly and may either decrease or increase slightly the coulomb efficiency or may have no significant effect on the coulomb efficiency (especially at pigment-to-binder ratios of 0.1/1 to 0.5/1).

THROWING POWER. The throwing power of an electrocoat primer is normally increased as the PVC content increases. This is usually the case at PVC levels from 1 to 45%. Pigments with high film resistance give the highest throwing power.

RUPTURE VOLTAGE. Increasing the PVC level from 1% to 15% in an electrocoat primer system usually has little effect on rupture voltage. Rupture voltage is increased markedly with increasing solids and also by vehicles which deposit a firm film of high viscosity.

GLOSS OF FILM. Smooth films with 60° gloss readings of 70° to 80° can be obtained at pigment-to-binder ratios up to 0.5/1. Generally at P/B of 1/1 or 1.5/1 rough textured films result. Tasker and Taylor (8) showed in iron oxide extender primers that the weight of resin deposited was constant, irrespective of the P/B level and that the weight of film deposited increased with increasing P/B ratio; therefore, more pigment was present with resulting lower gloss.

Effect of Bath Solids Level. Synthetic iron oxide maleinized oil electrodeposition primers have been successfully deposited at bath solids levels of 7.5 to 15%. No significant differences could be noted in dry film thickness or film hardness. Nonvolatile levels of 7.5 to 10% appear to be more satisfactory for general appearance of the film but slightly poorer than 12.5 to 15% levels for corrosion resistance properties. The normal production level is about 10% solids in the bath.

Effect of Specific Resistance of Iron Oxide Pigments. The effect of the soluble salt content and of the specific resistance of synthetic and natural iron oxides was evaluated in electrocoat primer systems. A high specific resistance and a low water-soluble salt content is significant to the efficiency of the electrocoat process only as it is related to a particular pigment. Washing the foreign electrolytes out of a pigment and increasing its specific resistance do not necessarily make this pigment perform more efficiently in an electrocoat system. Therefore, it is not the foreign electrolytes but those electrolytes which correspond to the solubility products of the pigment, its chemical composition, and its specific conductivity in aqueous suspensions that determine the efficiency of a particular pigment in an electrocoat system.

Although water-soluble salt content may not affect coating efficiency and deposited film integrity, a low soluble-salt content is desirable since an excessive accumulation of salts with repeated turnovers will cause poor tank stability and produce gases, resulting in films with poor integrity.

Effect on Resistance Characteristics of Deposited Film. The addition of pigment (synthetic red iron oxide) at PVC level of 1% to 15%, to an electrocoat primer system (maleinized oil type), generally will produce firmer, thicker films which results in improved corrosion resistance of the deposited electrocoat primer film. The addition of 2% of the total pigmentation as strontium chromate and/or 5 to 10% of the pigmentation as a basic lead pigment will significantly improve the salt fog resistance of an iron oxide electrocoat primer.

Selection of Pigments

Various selected grades of synthetic red and yellow iron oxides, chromium oxides, carbon and lampblack, titanium dioxide, phthalocyanine blue and green, and some organic yellows and reds have been successfully used in electrocoat paint systems, along with the following (*see also* Table II):

(1) Anti-corrosive pigments: various grades of lead pigments (basic white lead or silico chromate), chromates with low water solubility (strontium chromate), zinc sulfide and zinc oxide.

(2) Extender pigments: some grades of barytes, talc, calcium carbonate, kaolin, silicas, mica, and asbestos.

(3) Suspending pigments: beneficiated bentonite or hectorite clays, synthetic clays, and colloidal silicas.

Although raw material cost is still important in electrocoat formulations, the pigments selected must have good transfer and stability properties with the given vehicle system that is used. Many pigment manufacturers have done extensive work with their line of conventional pigments in various types of electrocoat vehicle systems; in many instances they have made tailored products to meet the electrocoat requirements. These pigments will need properties such as excellent stability in an alkaline media, low solubility in water, easy dispersibility, minimum settling, controlled particle size and shape (maximum of 6 microns), and a balance between specific resistance and specific conductivity in aqueous suspensions that allows good mobility or transfer properties.

Table II. Typical Formulation of Electrocoat Paints

	Ca. 100 gal Formula Parts by Weight
Red Oxide Automotive Primer	
Synthetic iron oxide	20
Basic white lead	2
Bentonite or hectorite clays	0.2
Amine solubilized maleinized oil vehicle	13
Triethylamine	1
Butyl Cellosolve	3
Deionized water	17
Disperse in porcelain or sand mill	
Reduction	
Amine solubilized maleinized oil vehicle	90
Diethanolamine	3
6% Manganese drier	0.2
Guaiacol	0.3
Deionized water	730
	880.0 lbs

Pigment volume concentration	5.0%
Non-volatile content	10.0%
pH	7.8—8.0
Weight per gal	8.8
Specific resistance, 1300 ohms	

Red or Black One-Coat System	
Carbon black or synthetic Iron Oxide	3.5
Water reducible epoxy–ester vehicle	13
Butyl Cellosolve	0.4
Triethylamine	0.4
Deionized water	17
Disperse in porcelain mill or sand mill	
Reduction	
Water reducible epoxy–ester vehicle	87
Butyl Cellosolve	2
Triethylamine	3
Deionized water	720
	846.3 lbs

Pigment volume concentration	1.0%
Non-volatile content	10.0%
pH	8.5
Weight per gal	8.5
Specific resistance, 1300 ohms	

Anti-Corrosion Pigments. Many of the regular anti-corrosion pigments used in conventional paint systems are too soluble to be used in electrocoat systems. This is especially true of most of the chromate pigments. Their relatively high solubility in an aqueous electrocoat tank where electrolysis is taking place may cause precipitation of the metal cations and release of free chromate ions into the bath; these ions can react with the electrocoat vehicle and cause flocculation problems. Low solubility lead and zinc pigments along with strontium chromate appear to work best in most electrocoat vehicle systems. However each vehicle system will have different properties, and each individual anti-corrosion pigment must be evaluated for stability in each specific vehicle system that is used.

Pigment Formulation Variables. Decreasing the pigment volume concentration of an electrocoat paint or primer system will usually give a softer film with greater film smoothness but less hiding power. Increasing the pigment volume concentration will give harder films which are rougher in appearance. This increase in PVC will usually increase the throwing power and also make the deposited film more susceptible to pinholing.

Prime, extender, and anti-corrosion pigments have a definite effect on the stability of the electrocoat paint system. Pigments which have the desired stability in alkaline medias, proper size, chemical composition, and low solubility with accompanied high specific resistance have been used successfully at PVC levels of 1 to 20%. Usually the type of vehicle has more effect on the stability of the electrocoat system than does the pigment, but it is still important that each pigment or combination of pigments be thoroughly evaluated (at the desired PVC level) in the laboratory before large scale production tank trials are made. Certainly the increased use of a pigment whose electrolyte content is of the type that contributes to the pigment's high solubility in water will undoubtedly cause flocculation and tank stability problems.

Testing Methods for Performance

Electrocoat paints and/or tanks are generally tested for the following properties:

(1) *Throwing power:* measuring the percent of paint deposited inside a ¾-inch standard gas conduit (with or without a steel panel inside) or the percent paint deposited on the inside of either two or three phosphated steel panels that are separated by ⅜-inch shims.

(2) *pH and alkalinity:* 50 ml of the paint are usually checked with a pH meter. The alkalinity is usually determined in milliequivalents of amine per 100 grams of paint solids.

(3) *Conductivity:* may be measured with a circulation cell connected to a measuring bridge of the Wheatstone type, the paint being the unknown resistance.

(4) *Non-volatile content.*

(5) *Thickness of the cured film:* may be measured in mils by electronic thickness tester or in microns by a Permascope.

(6) *Ash-binder ratio:* 10 to 15 mil heated at 220°F to determine non-volatile content. The non-volatile is then ashed at 1400°F for one-half hour. This ratio is very important because one can determine the ash-to-binder ratio of the tank and also that of the deposited film. Normally the pigment will deposit more efficiently than the binder, and the pigment-to-binder ratio will decrease in the tank as it is worked. Adjustments must be made to keep this ratio in balance.

(7) *Coulombic yield:* usually expressed as the weight in milligrams deposited on the anode for each coulomb consumed and is calculated from a recording amperemeter.

Other laboratory tests may be performed such as pumping stability and continuous process testing of the life span of an electrocoat bath. A coil stock coating device with variable speed drive, heat exchangers, and paint feed equipment is used. Feed, temperature, and throughput are run the same as in a production tank. The tank is run continuously until the deposited films do not meet the specifications or usually until 20 times the weight of solids used as original fill has been converted into coating. A 5-gallon laboratory tank will completely turn over in one day whereas it takes about 20 days for a production tank to turn over. With the present knowledge of electrocoat systems and by using ultrafiltration the bath life of an electrocoat automotive primer tank can be controlled for indeterminate periods of time.

Literature Cited

1. "The Influence of the Construction of Electrocoat Installation on Coating Results," Dr. H. Frangen, Conference on Electro-Painting, London 28 and 29, October, 1969, organized by Business Conferences & Exhibitors, Ltd.
2. Levinson, S. B., "Electrocoat, Powdercoat, Radiate," *J. Paint Technol.* (June 1972) **44**, 569; Pt. I p. 49.
3. Hauser, Ernest A., "Latex," translated by W. J. Kelly, pp. 136–137, Chemical Catalog Co., New York, 1930.
4. Feinlieb, M., "Electrodeposition of Vinyl Plastics," *Trans. Electrochem. Soc.* (1945) **88**, 11.
5. Mosley, J. R., Wallace, T. C., "Electrophoretic Deposition, A Versatile Coating Method," *J. Electrochem. Soc.* (1962) **109**, 923.
6. Pearlstein, F., Wick, R., Gallaccio, A., "Electrophoretic Deposition of Metals," *J. Electrochem. Soc.* (1963) **110**, 843.
7. Berg, D., Hirayama, C., "Studies on the Electrophoretic Deposition of Mica," *Electrochem. Technol.* (1963) **1**, 224.
8. Tasker, L., Taylor, J. R., *J. Oil Colour Chemist's Assoc.* (1965) **48**, 121.

RECEIVED May 28, 1971.

8

Studies in Cathodic Electrodeposition

R. A. WESSLING, D. S. GIBBS, W. J. SETTINERI, and E. H. WAGENER

Physical Research Laboratory and Designed Products Department, The Dow Chemical Co., Midland, Mich. 48640

> *Coatings have been electrodeposited on a steel cathode from an alkaline bath containing a cationic sulfonium latex. As with quaternary ammonium stabilized latexes, the charge on the polymer particles is independent of pH, permitting operation on the alkaline side. In a comparative study, however, only the sulfonium latex deposited a smooth, thin (< 1 mil) coating with rapid current cut off and low residual current. The mechanism of anodic electrodeposition was reviewed and compared with the results of this study. The protonated amine systems appear to destabilize at the electrode by a neutralization mechanism, in analogy with the neutralization of carboxyl groups in the anodic process, but this mechanism is not available to the sulfonium system. Therefore, a unique mechanism of deposition must be operating in the latter case.*

The process of electrodepositing organic coatings on metal cathodes has been investigated. Cathodic electrodeposition has a number of inherent advantages over the commonly used anodic processes, but these are best realized if the process can be carried out in a neutral or alkaline medium. A review of the literature suggests that the major problem in developing cathodic electrodeposition has been the lack of cationic emulsions that are stable above pH 7 and still deposit electrically insulating films on the cathode surface.

An investigation of the chemistry of sulfonium salts in our laboratories suggested that emulsions stabilized by these groups might have the desired characteristics. In recent years, the capability of preparing such emulsions has been achieved (1). This paper gives a preliminary account of their performance in electrodeposition and provides further insight into the mechanism of deposition at the cathode. The results suggest that significant differences exist between anodic and cathodic processes.

Electrodeposition of organic coatings onto metal surfaces has been known for about 40 years, but not until the early 1960's has it come into industrial prominence, largely because of the efforts of Brewer, Burnside, and co-workers at Ford Motor Co. (2). At present, use of this process is growing rapidly as evidenced by the number of installations and patents relating to the process (3, 4).

Review of Anodic Systems

To appreciate the potential advantages of a cathodic system, particularly an alkaline cathodic system, a brief review of the mechanism of anodic electrodeposition is helpful. The anodic systems developed to date have been generally based on the incorporation of the carboxylate anion (R–COO$^-$) either as a part of the polymer molecule itself or the surfactant of an emulsion. Many studies have been made on the mechanism of electrodeposition involving carboxylate stabilized systems (5–11), and it is generally agreed that the following reactions contribute to the formation of an electrically insulating film on the anode:

(1) Electrolysis of water:
    (a) $H_2O \rightarrow OH\cdot + H^+\, e^-$
    (b) $2\, OH\cdot \rightarrow H_2O + O\cdot$
    (c) $2\, O\cdot \rightarrow O_2$

(2) Oxidation of the anode:
    $M^\circ \rightarrow M^{n+} + ne^-$

(3) Reaction of the carboxylate ion with H^+ and M^{n+}

$$R-\underset{\underset{O}{\|}}{C}-O^- + H^+ \rightarrow R-\underset{\underset{O}{\|}}{C}-OH$$

$$*R-\underset{\underset{O}{\|}}{C}-O^- + M^{n+} \rightarrow (R\underset{\underset{O}{\|}}{C}-O)_n-M$$

In addition, Berry (11) has suggested that concentration coagulation could occur simultaneously with the carboxylate reactions shown above. This mechanism would allow some of the carboxylate salt to be deposited in the film intact.

This is a simplified view of the anodic deposition mechanism which leaves out many parameter effects studied in detail by workers in the field, but it basically describes most anodic systems in that the deposited films appear to contain the products predicted by these reactions. Accepting this mechanism as the basic model for anodic electrodeposition, several disadvantages become immediately evident.

(1) The system is highly pH dependent. Any lowering of the pH significantly below 7 would coagulate the system since stabilization depends on the carboxylate ion's being present. High pH's have been shown to produce undesirable side effects such as gassing. Most systems operate in a pH range of 7 to 8.7. In addition, the electrodeposition performance is sensitive to changes in pH since the carboxylate ion concentration is pH dependent.

(2) The free carboxyl group produced by reaction with H^+ remains in the coating and is usually not decomposed during the baking step. The ionizable group presents a point of vulnerability to moisture transmission through the coating.

(3) Metal ions are included in the film. In the case of a copper substrate, this has led to green coatings (12). Coatings on steel may be similarly discolored (13). The metal carboxylates or metal hydroxides (10) formed in the coating are points of vulnerability for moisture transmission. The metal carboxylates would be susceptible to hydrolysis, which some workers believe could result in corrosion at the coating–metal interface (14).

(4) The electrolysis of water at the anode produces oxygen. The generally accepted mechanism of its formation involves both the hydroxyl radical and the oxygen radical. Some workers (15) report that reactions of these species with the coating have an adverse effect on the coating performance properties and even change the resin character in solution. The oleoresinous systems seem to be particularly affected and require antioxidants.

(5) In general, relatively large amounts of base are required to create the carboxylate anion. If amines are used, they generally require from about 0.3 to 0.5 milliequivalents/gram of polymer solids to produce acceptable performance.

These disadvantages have obviously not stopped the growth of anodic electrodeposition. However, it is interesting that the theoretical solutions to some of these problems have been known to be obtainable by cathodic deposition, yet no commercial cathodic system has appeared.

Review of Cathodic Systems

Several advantages of cathodic electrodeposition make it a worthwhile area of investigation. Reduction takes place at the cathode; therefore, the metal surface does not ionize and remains passive during the deposition process. No ions enter the coating even with easily oxidizable substrates. The electrolysis of water at the cathode produces the hydrogen radical, hydrogen, and hydroxide ion.

$$2 H_2O + 2 e^- \rightarrow 2 OH^- + 2 H \cdot$$
$$2 H \cdot \rightarrow H_2$$

Reaction of these species with the coating material will probably have a less harmful effect on coating properties than the corresponding oxidation reaction mentioned above.

Most of the cathodic systems described in the literature are simple analogs of the anodic resins where the carboxylate group ($-COO^-$) has been replaced by an amino group ($-\overset{|}{\underset{|}{\overset{+}{N}}}-H$). These resins are prepared in a nonaqueous reaction and post emulsified in water to form an electrocoating system. They generally have a charge/mass ratio high enough to make the polymer water soluble or at least water dispersable. Conventional latexes made by emulsion polymerization are not commonly used. A recent broad review (16) on the chemistry and technology of cationic polyelectrolytes focused on quaternary ammonium polyelectrolytes but also described other types of polyelectrolytes such as sulfonium and phosphonium. The long list of industrial applications identified in this paper did not include electrocoating, indicating that onium ion polyelectrolytes suitable for this use had not been developed. Two important applications for quaternary polyelectrolytes, electroconductive coatings and antistatic agents, require the polymer to retain charge for performance. Electrocoating usually requires a rapid destruction of charge in order to optimize throwing power, current efficiency, reduction of gassing, etc. This suggests that the quaternary ammonium group and electrodeposition are not very compatible.

In comparison with the voluminous literature on quaternary ammonium polyelectrolytes and surfactants, the sulfonium analogs have received little study. The review cited above introduced sulfonium polyelectrolytes with the statement "the sulfonium quaternary unit does not seem to have any special efficacy over the more readily available ammonium quaternary group." It also mentions their chemical instability (compared with quaternary systems) as a disadvantage. Jungermann (17), in his book on cationic surfactants comments: "It is difficult to imagine what advantages these onium (non-nitrogeneous-surfactants) compounds would possess over the corresponding ammonium compound except perhaps in the area of biological activity."

These comments indicate a common attitude toward the use of sulfonium salts in polymer technology. Nonetheless, as Hatch pointed out (18), the instability and high reactivity of sulfonium salts are an advantage in certain cases. As conventional coatings for example, sulfonium-based systems including both polyelectrolytes (19, 20) and latexes (21, 22) can be deposited from aqueous media and easily cured to hydrophobic products.

The results of our study indicate that the higher chemical reactivity of sulfonium compounds is also an advantage in electrodeposition of organic coatings, but heretofore such systems have not been reported.

The cationic electrodeposition systems which have been described in the literature include both soluble systems and emulsions, but all have been based on nitrogen cations. Two types have been reported:

(1) Protonated amine based systems.

$$-\overset{|}{\underset{|}{N^+}}-H$$

(2) Quaternary ammonium systems.

$$(R)_4{}^+-N$$

Work on the first type of system is reported in a series of patents by Spoor, Pohlemann, and co-workers (23–25) in which they describe electrodeposition of both solutions and dispersions having a $-\overset{|}{\underset{|}{N^+}}-H$ cation as the attractive vehicle. In general, they employ aminoalkyl esters of acrylic acid such as mono(N,N-dimethylamino)ethyl methacrylate(I) or N-vinylimidazole(II) as comonomers in their polymeric compositions. The cations are formed by adjusting the pH with either acetic or hydrochloric acid to give solutions (or opalescent solutions) which are subsequently electrocoated on the cathode. Their general conditions for electrodeposition are:

pH	3.5–6.0
Applied voltage	20–100 volts
Solids	8–10%
Electrodeposition time	2 minutes

They report coatings ranging from 0.46–1.32 mils thick after baking which have resistance of varying degrees to salt water and alkaline water.

Slater and Thow (26) describe electrocoating of epoxy ester dispersions using salted amine dispersing agents, all of which are the $-\overset{|}{\underset{|}{N^+}}-H$ type. In their examples, they describe dispersions made with dimethyl soya amine, 1-hydroxyethyl-s-heptadecenyl imidazoline and dimethyl hydrogenated tallow amine. They claim organic acids of varying molecular weights to form the amine cations. Formic acid is used in their reported examples.

General conditions for electrodeposition are:

pH	3.4–4.2
Applied voltage	200–250 volts

Solids	10%
Electrodeposition time	Continuous coil coating

The resultant coatings are 1.0–1.1 mils thick on steel cathodes.

Tawn (27) reported a nitrogen-based cationic system which comprises a dispersion of an epoxy resin with a "partially neutralized aminoamide or imidazoline (which may or may not be polymeric) or with an epoxy-adduct thereof." He claims several novel advantages, such as rapid deposition at low voltages (20–500), high current efficiencies, very dilute solutions (1% or less), and coatings which cure at room temperature or slightly above. No data are given, but the system probably operates at a pH < 7 based on the $-\overset{|}{\underset{|}{N^+}}-H$ structures mentioned.

Four additional patents on cathodic electrodeposition have been recently issued (28–30), but the techniques used in these cases to get cathodic deposition are essentially the same as that described above. These patents differ primarily in resin composition and post-curing mechanisms.

All of the above mentioned systems are based on amine salts. McCoy (31) has reported the only case of electrodeposition of quaternary ammonium-based emulsions. Three specific derivatives he mentioned are: N-dodecylbenzyl-N,N-diethyl-1-ethanol ammonium chloride; 1-(2-aminoethyl)-2-1-alkenyl-2-imidazole; and octadecenylmethyl di-2-hydroxyethylammonium chloride.

Asphalt emulsions stabilized with these emulsifiers were electrodeposited under the following conditions:

pH	3.0–6.0
Applied voltage	30–100 volts
Solids	10–30%
Electrodeposition time	Varied

The coatings deposited were characterized by weight/sq inch with no thickness given. The data were presented to show that cationic emulsions deposit coatings at a higher rate and a higher efficiency than anionic emulsions. The weights of asphalt deposited were generally in the range of 200 mg/sq inch, which is relatively high compared with the results obtained from the sulfonium emulsions used in this study. Further treatment of McCoy's data is presented in the discussion section.

To the authors' knowledge, none of these cationic systems is of significant commercial importance today. The remainder of this paper demonstrates some of the problems that might be encountered with the systems above and shows that sulfonium-based systems can overcome these difficulties.

Experimental

The majority of the experiments were conducted in a polyethylene cell 10.7 cm × 7.9 cm × 2.54 cm. Two, equal size graphite anodes were placed at the ends of the long axis of the cell. A 10.2 cm × 1.27 cm metal sample was placed equidistant between the graphite anodes (usually 4.6 cm) so that the flat plane of the sample was normal to a line joining the two anodes. The power source was capable of producing three-phase direct current at 500 volts and 40 amps. A Bausch and Lomb VOM-5 recorder was used to plot current as a function of time. The 20 inches/minute speed was used for all experiments to expand the current–time profile best. Occasionally, a second VOM-5 recorder was used in conjunction with a standard calomel electrode to plot the voltage drop between the cathode and the solution as a function of time. The area under the current–time curve was integrated using an Infotronics CRS-110 integrator provided with a Victor Digit-matic data printer. A Microflex timer was placed in the circuit so that time intervals as short as 1/10 of a second up to 2 minutes could be accurately reproduced. The system was automated so that a button activated the timer which allowed a preset voltage to be applied across the cell for a precise interval, at the end of which the integrated area (coulombs) was printed out. Elaborate safety controls were provided for the system. Data on coulombs, coating weight, and current efficiencies represent the average of four experiments. The metal used for this study was steel with Bonderite 37 coating. Each sample was accurately cut so that the area of the sample coated varied no more than 1%. For small coupons this area was 7.3 sq cm.

In a typical experiment, 70 grams of material at 10% solids were placed in the cell containing the metal sample. The power source was preset for 200 volts, and the experiment was activated as described above. At the end of 2 minutes, the sample was removed from the solution and rinsed thoroughly with deionized water. The sample was air dried and weighed. The current efficiency was determined for each sample from the weight of polymer deposited per coulomb (c) passed by the following equation:

$$\text{Current efficiency} = \frac{\text{Weight (mg)}}{\text{coulombs}}$$

Percent solids (total weight basis), pH, conductivity, stability of the latexes, and emulsifier coverage were determined by standard techniques. Particle size was determined by light scattering (Brice-Phoenix) and was occasionally checked by electron microscopy.

Results

To supplement the limited experimental data published on the nitrogen-based systems, several latex and water-soluble systems were synthesized and characterized. The —$\overset{|}{\overset{+}{N}}$—H literature was somewhat vague about the colloidal nature of their systems, referring to them as "dispersions" or "opalescent solutions." In this work, a latex system with well defined colloidal properties was synthesized using a $-\overset{|}{\underset{|}{N^+}}-H$ emulsifier (Figure 1). The latex characteristics are also given.

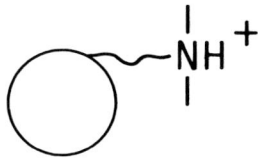

Figure 1. Schematic of an $-\overset{|}{\underset{|}{N^+}}-H$ stabilized latex

Latex characteristics:
    Composition   60 butyl acrylate/40 styrene
    Emulsifier concentration   0.10 meq/gram polymer
    Particle size   1100 A
    Solids   10%
    pH   2.5

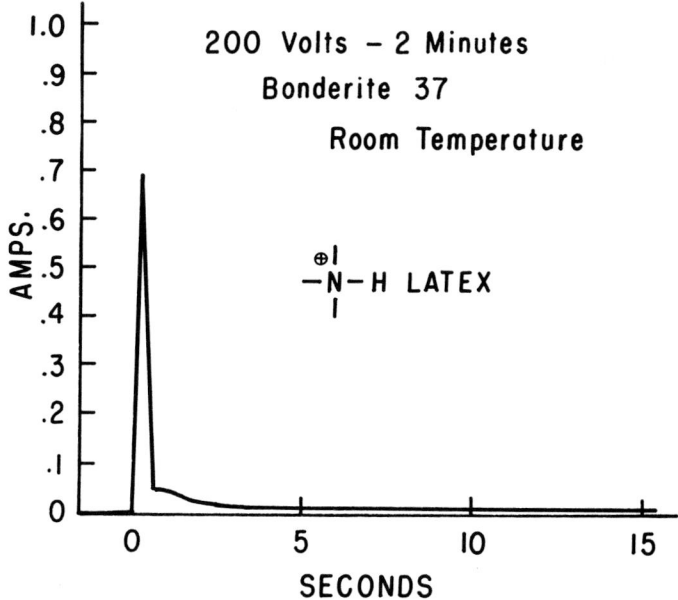

Figure 2. Current vs. time for $-\overset{|}{\underset{|}{{}^+N}}-H$ stabilized latex

The current–time curve for this system is shown in Figure 2. The curve shows a rapid cutoff of current and low residual current which indicates that the system would have good throwing power (32). The electrodeposition performance data are shown in Table I.

Table I. —N⁺—H Latex Electrodeposition Performance

Coating weight	0.97 mg/sq cm
Coating thickness	0.6 mil
Coating appearance	smooth, uniform
Coating efficiency	8.3 mg/c
Residual current	0.14 ma/sq cm

As noted earlier, these experiments were run at a pH of 2.5. Their performance couldn't be evaluated above a pH of 7 owing to coagulation of the latex.

Two quaternary ammonium systems were synthesized: a water-soluble polyelectrolyte and a latex (Figures 3 and 4). The polymer composition and onium ion structure were made as nearly alike as possible to avoid effects from structural changes. The two systems differed significantly in the number of charges per unit mass of polymer as would be expected in comparing a soluble polyelectrolyte and a latex. The water-soluble system was electrodeposited at 5% solids and was stabilized with 2.1 meq of the quaternary moiety per gram of polymer. The resultant current–time curve is shown in Figure 5. As the curve indicates, the water-soluble system remains conductive and deposits only a gel-like coating, rough and bubbly in appearance and generally unacceptable as

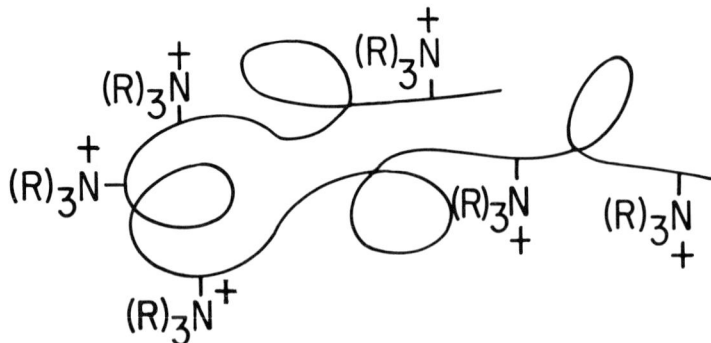

Figure 3. Schematic of an $(R)_4N^+$ stabilized water-soluble polymer

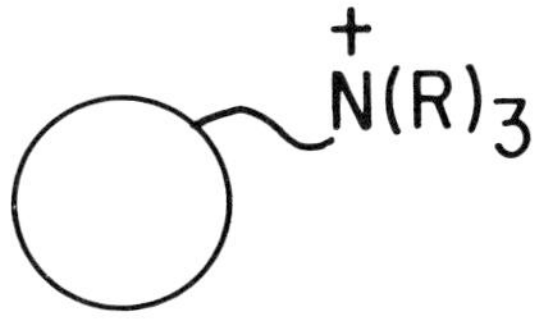

Figure 4. Schematic of an $(R)_4N^+$ stabilized latex

Latex characteristics:
Composition   60 butyl acrylate/40 styrene
Emulsifier concentration   0.04 meq/gram polymer
Particle size   1100 A
Solids   10%
pH   7.7

a metal coating. The bath temperature rose rapidly as the current continued to flow. Varying solids and voltage did very little to give a better coating from this system.

The quaternary ammonium latex, on the other hand, produced a significantly better current–time curve as shown in Figure 5. The current did drop, but the electrodeposition performance of the quaternary ammonium latex was not as good as the electrodeposition performance of the —$\overset{|}{\underset{|}{N^+}}$—H latex as shown in Table II.

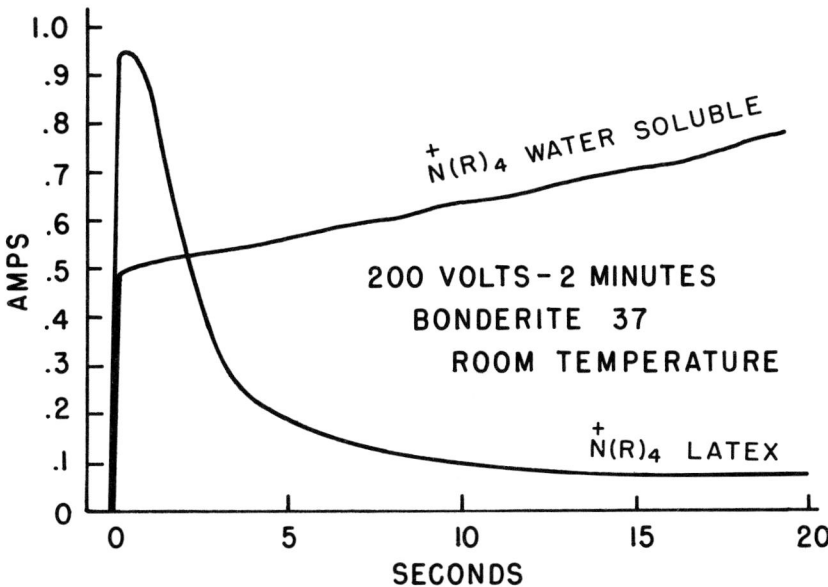

Figure 5. Current–time curves of water-soluble and emulsion quaternary ammonium systems

Table II. Electrodeposition Performance for $-\overset{+}{\text{N}}(\text{R})_3$ and $-\overset{+}{\text{N}}-\text{H}$ Latexes

	$-\overset{+}{\text{N}}(\text{R})_3$	$-\overset{+}{\text{N}}-\text{H}$
Coating weight	7.3 mg/sq cm	0.97 mg/sq cm
Coating thickness	4.0 mils	0.6
Coating appearance	rough, bubbly	smooth, uniform
Current efficiency	42 mg/c	8.3 mg/c
Residual current	0.82 ma/sq cm	0.14 ma/sq cm

Since the quaternary ammonium systems are fully ionized, the stability of these latexes is more independent of pH. The data comparing the electrodeposition performance in Table II should be qualified on two points: the pH of the quaternary system was 7.7 vs. 2.5 for the $-\text{N}^+-\text{H}$ latex, and the emulsifier concentration was higher (0.1 meq/gram vs. 0.04 meq/gram) for the $-\text{N}^+-\text{H}$ system. However, the performance of the quaternary ammonium latex system did not improve at lower pH or higher emulsifier concentration. The $-\text{N}^+-\text{H}$ system would probably perform even better with respect to current efficiency if the emulsifier concentration were lowered to 0.04 meq/gram. Therefore, the conclusion that the $-\text{N}^+-\text{H}$ system performs better with respect to desirable coating thickness, rapid current cutoff, low residual amperage, film appearance, and coating weight remains valid.

In general, the data on the electrodeposition performance of the nitrogen based systems can be summed up as follows:

System	Electrodeposition Performance
$-\text{N}^+-\text{H}$ Solution	Good
$-\text{N}^+-\text{H}$ Latex	Good
$(\text{R})_4\overset{+}{\text{N}}$ Solution	Unacceptable
$(\text{R})_4\overset{+}{\text{N}}$ Latex	Fair

As discussed later, the differences between the $-\overset{|}{\underset{|}{N}}{}^{+}-H$ solution and $^+N(R)_4$ solution performance is probably caused by the lack of a simple charge destruction mechanism:

$$-\overset{|}{\underset{|}{N}}\overset{+}{H} + OH^- \rightarrow -\overset{|}{\underset{|}{N}} + H_2O$$

for the $N^+(R)_4$ solution. Better performance of the $N^+(R)_4$ latex over the $N^+(R)_4$ solution can be attributed to the fact that the latex can undergo destabilization (concentration coagulation) to deposit a film whereas a water-soluble polymer would not readily coagulate.

A sulfonium polyelectrolyte (essentially the same as the previously described quaternary ammonium polyelectrolyte except for the onium ion structure) stabilized with 2.3 meq of the sulfonium moiety per gram of polymer was synthesized and subjected to the same electrodeposition conditions described earlier for the quaternary ammonium polyelectrolyte. The performance under these conditions was not significantly better than that of the quaternary ammonium polyelectrolyte. Therefore, attention was centered on the sulfonium latex systems which gave superior coatings.

The sulfonium latex was comparable with the quaternary ammonium latex above with respect to polymer composition, particle size, emulsifier structure, emulsifier concentration, and polymer concentration. The difference again was in the onium ion structure. The latex characteristics, current–time curves, and electrodeposition performance are shown in Figure 6. Note that the sulfonium system was run under alkaline conditions at a pH of 7.6.

Under the identical electrodeposition conditions used for the quaternary ammonium stabilized latex, the sulfonium stabilized latex de-

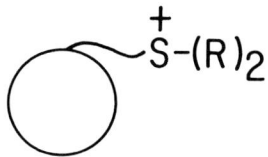

Figure 6. Schematic of an $(R)_3S^+$ stabilized latex

Latex characteristics:
  Composition   60 butyl acrylate/40 styrene
  Emulsifier concentration 0.04 meq/gram polymer
  Particle size   1100 A
  Solids   10%
  pH   7.6

posited a film with better throwing power (as indicated by the slope of the curves) (32), lower residual current, and better appearance. In addition, less heat was generated in the bath than with the quaternary ammonium latex. The film was smooth and uniform with a thickness in the 1.0 mil range. The comparative results are shown in Figure 7 and Table III.

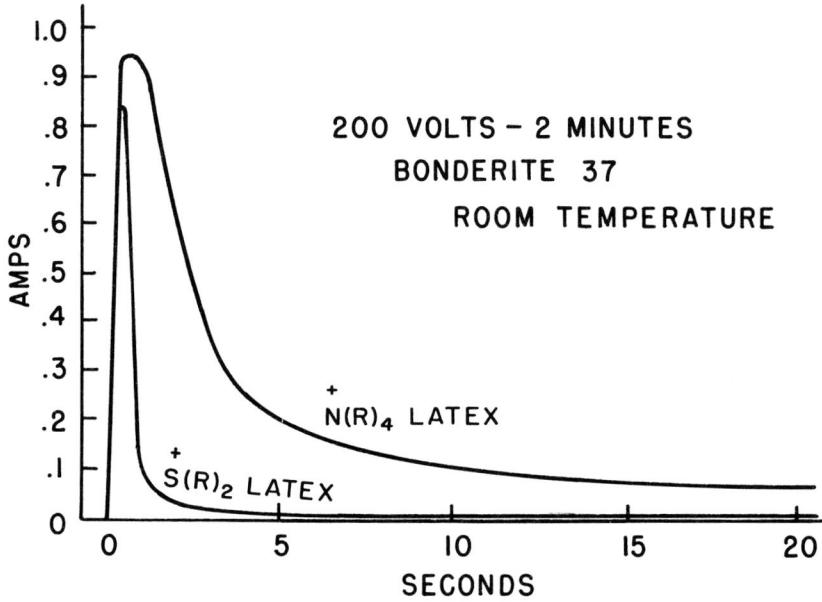

Figure 7. Current–time curves for sulfonium and quaternary ammonium latexes

Table III. Electrodeposition Performance for —$S^+(R)_2$ and —$N^+(R)_3$ Latexes

	—$\overset{+}{S}(R)_2$	—$\overset{+}{N}(R)_3$
Coating weight	1.6 mg/sq cm	7.3 mg/sq cm
Coating thickness	0.8 mil	4.0 mils
Coating appearance	smooth, uniform	rough, bubbly
Current efficiency	44 mg/c	42 mg/c
Residual current	0.09 ma/sq cm	0.82 ma/sq cm

Discussion

Before discussing the quaternary ammonium *vs.* sulfonium latexes, some comments should be made about the utility of the —$\overset{|}{\underset{|}{N^+}}$—H systems.

At first glance they are attractive systems since both solutions and latex deposit good films with good electrodeposition performance. Some of these systems are being field evaluated although no commercial use has been adopted. The reason probably lies in the lack of a clear advantage over the anodic systems. The data of Wismer and Bosso (33) indicate that the postulated inherent cathodic advantages of no oxidation of the substrate (no metal ions in the film) and no oxidation of the film are valid. However, these advantages may be outweighed by the fact that these systems must operate at an acid pH where corrosion of the equipment and possibly the metal being coated would be a problem. In addition, the performance of these systems, like the performance of anodic carboxylate (COO^-) systems, would be very pH sensitive.

The big advantage of the quaternary ammonium and sulfonium systems over the $-\overset{|}{\underset{|}{N^+}}-H$ systems is less sensitivity to pH. The quaternary salts, being strong electrolytes, can be used above pH 7. Their high degree of ionization results in the cation concentration being more constant during pH changes in the bath. This should be reflected in more stable electrodeposition performance. These advantages, when added to the inherent cathodic process advantages, make the quaternary systems worth investigating. Yet, of the quaternary ammonium and sulfonium systems investigated, only the sulfonium latex performed well enough to compete with the commercial anodic systems.

As mentioned, one possible reason for the poor behavior of the quaternary ammonium system is that the simple acid–base charge destruction mechanism of the $-\overset{|}{\underset{|}{^+N}}-H$ and anodic systems is no longer available. In the anodic systems, protons are produced by the electrolysis of water, which reacts with the incoming carboxylate stabilized resin to form a water insoluble product.

$$———\overset{O}{\underset{}{\overset{||}{C}}}-O^- + H^+ \rightarrow ———\overset{O}{\underset{}{\overset{||}{C}}}-OH$$

Water soluble   Water insoluble

Since hydroxyl ions are generated at the cathode, a similar mechanism could be envisioned for the $-\overset{|}{\underset{|}{^+N}}-H$ systems, where

$$———\overset{|}{\underset{|}{N^+}}-H + {}^-OH \rightarrow ———\overset{|}{\underset{|}{N}} + H_2O$$

the charge is destroyed by the acid–base reaction above. This mechanism is available to both amine polyelectrolytes and amine-stabilized emulsions. The quaternary ammonium systems, on the other hand, are not destabilized by pH changes. Therefore, a different mechanism is required to get deposition of a non-conductive polymer out of a quaternary ammonium stabilized system.

Our results show that quaternary ammonium polyelectrolytes deposit conductive, water swollen gels. This correlates with their performance as electroconductive resins (16), which is based on the retention of the R_4N^+ group in the dry polymer. In our studies, the fact that residual quaternary ammonium groups remain in the coating was confirmed by mass spectral analysis.

Since the water-soluble quaternary ammonium polyelectrolytes are lyophilic colloids, they can be irreversibly coagulated only by destroying the charged groups. Apparently this does not occur in the electrodeposition process of quaternary ammonium, so the polyelectrolyte is in effect only concentrated at the electrode under the influence of the applied electric field. It is unlikely that any throwing power could be achieved with such a system.

The quaternary ammonium stabilized latex, however, is a lyophobic colloid. Therefore, the mechanisms of concentration coagulation and electrolyte coagulation are available to destabilize the system. Under the conditions used in our experiments, electrolyte coagulation seems the less likely of the two. Hydroxide ions are generated at the electrode surface but probably not in sufficient quantity to flocculate the latex particles at the interface.

Concentration coagulation is the better possibility. Both the sulfonium and quaternary ammonium can be irreversibly coagulated simply by drying. A cast layer of latex quickly coalesces into a continuous, water-insoluble film as the water is removed. The forces bringing about coalescence at the electrode surface are of course quite different, but the same general principle applies. If the latex particles are brought close enough together, the dispersion forces cause irreversible coalescence, but to do this, the electrostatic repulsion between the charged particles must be overcome. In electrodeposition, this is done by the process of electro-endo-osmosis.

The above quaternary ammonium latex system, even though it does have some throwing power because of destabilization, deposits a relatively thick coating which is rough and bubbly in appearance. In the current study, 7.3 mg/sq cm of polymer were deposited. This coating was 4.0 mils thick. Additional data showing that quaternary ammonium systems deposit thick coatings comes from the study of McCoy (31) mentioned earlier. Using N-dodecylbenzyl-N,N-diethyl-N-ethanol ammonium chloride as the emulsifier, he reports the following:

pH	Electrodeposition Time, minutes	Asphalt Deposited, mg/sq in./min	Asphalt Deposited, mg/sq in.	(mg/sq cm)
5.4	1.0	82	82	(11.2)
5.0	2.0	70	140	(19.2)
4.5	5.0	48	240	(32.8)
3.0	3.0	67	201	(27.5)

These weights are considerably higher per square inch than those found in this study. In still another example, he reports that an emulsion stabilized with octadecenyl methyl di-2-hydroxyethylammonium chloride deposited 2,660 mg/sq inch after 30 minutes. Apparently the quaternary ammonium latex systems retain enough charge in the depositing film long enough to allow the film to build up to relatively thick coatings which would be unacceptable for most electrodeposition applications.

Even if these thick coatings were acceptable, it is doubtful that these films would have the desired water and corrosion resistance. The residual quaternary ammonium groups would allow water to permeate the coating even faster than residual carboxyl or amine groups owing to their high ionic character. Since they are relatively stable, thermal decomposition of the residual quaternary ammonium groups would be difficult.

As shown in Figure 3 and Table III, the sulfonium latex gave an excellent coating with good electrodeposition performance. The difference in performance between the sulfonium and quaternary ammonium latexes could be attributed to the higher chemical reactivity of the sulfonium group which causes more rapid emulsifier group destabilization resulting in more rapid coagulation of the latex.

Several mechanisms for charge destruction for quaternary type systems could be involved during electrodeposition. Most favor the sulfonium group's reacting at a higher rate. Investigations into these mechanisms are continuing and will be reported in subsequent papers. Initial results indicate that several mechanisms may be operating either concurrently or independently in sequence during the very short time it takes to deposit a coating.

Conclusions

A new approach to electrocoating has been found which involves cationic electrodeposition of an organic coating over a wide pH range. This approach involves the use of a sulfonium system which, under equal conditions, outperforms the quaternary ammonium systems reported previously. A different electrodeposition mechanism is probably involved from that suggested for the anodic systems and the $-\overset{|}{\underset{|}{N^+}}-H$ cathodic systems.

The cathodic sulfonium system has many advantages over the commercial anodic systems:

(a) It is a latex system which allows well-characterized colloidal properties and polymer composition. Finn and Mell (34) showed that the electrical current efficiency for latex systems is potentially significantly higher than water-soluble systems.

(b) It has wide pH range operability—good electrodeposition performance has been obtained from pH 2–10. In general, electrical efficiency increases with increasing pH.

(c) Sulfonium groups are easily thermally decomposed (90°–125°C) to give hydrophobic products, leaving no ionic groups in the film (19, 21, 22).

(d) In the sulfonium latex system, fewer sulfonium groups per unit polymer mass are needed for electrodeposition than the number of anionic groups used in the anodic system.

(e) No metal ions are formed at the cathode to discolor the film or increase the corrosion rate of the metal substrate. The latter effect has recently been demonstrated by May (14) in anodic electrodepositions.

(f) No oxidation of the film occurs since hydrogen is produced at the cathode.

(h) Early indications are that the sulfonium systems run cooler than the anodic system.

(i) This cationic electrodeposition process is applicable to a variety of polymer compositions.

Acknowledgments

The authors are especially indebted to H. D. Clarey and L. D. Yats for their excellent supporting work in these studies. The authors also gratefully acknowledge the technical assistance of B. W. Miller and J. L. Townsend and the stimulating conversations with T. Alfrey, Jr.

Literature Cited

1. Gibbs, D. S., Wessling, R. Q., Wagener, E. H., Belgian Patent **741,085** (1970).
2. Brewer, G. E. F., Burnside, G. L., U.S. Patent **3,355,373** (1967).
3. Phillips, G., Electroplating for the 70's Conference, Paper No. 15, London, England, Oct. 28-29 (1969).
4. Kiefer, D. M., *Chem. Eng. News* (1969) **47**, 32.
5. Hays, D. R., White, C. S., *J. Paint Technol.* (1969) **41**, 461.
6. Nakamura, Y., Oki, F., Nozaki, A., *Bull. Chem. Soc. Japan* (1969) **42**, 1534.
7. Rheineck, A. E., Usmani, A. M., *J. Paint Technol.* (1969) **41**, 597.
8. Brushwell, W., *Australian Paint J.* (1969) **18**.
9. May, C. A., Smith, G., *J. Paint Technol.* (1968) **40**, 493.
10. Mercouris, S., Graydon, W. F., *J. Electrochem. Soc.* (1970) **1117**, 717.
11. Berry, J. R., *Paint Tech.* (1963) **27**, 13; (1964) **28**, 24.
12. Hutchinson, C. O., *Plating* (1967) **54**, 1246.
13. Deibert, R. J., *J. Paint Technol.* (1966) **38**, 421.
14. May, C. A., *J. Paint Technol.* (1971) **43**, 43.

15. Sullivan, M. R., *J. Paint Technol.* (1966) **38**, 424.
16. Hoover, M. F., *J. Macromol. Sci. (Chem.)* (1970) 1327.
17. Jungermann, E., "Cationic Surfactants," p. 192, Marcel Dekker, New York, 1970.
18. Hatch, M. J., *Chem. Eng. News* (1960) **38**, 104.
19. Hatch, M. J., Meyer, F. J., Lloyd, W. D., *J. Appl. Polymer Sci.* (1969) **3**, 721.
20. Fang, J. C., U.S. Patent **3,310,540** (1967).
21. Kangas, D. A., U.S. Patent **3,322,737** (1967).
22. Lloyd, W. G., U.S. Patent **3,409,660** (1968).
23. Spoor, H., Florus, G., Pohlemann, H., Schauder, F., U.S. Patent **3,455,806** (1969).
24. Spoor, H., Florus, G., Pohlemann, H., Schauder, F., U.S. Patent **3,454,482** (1969).
25. Spoor, H., Pohlemann, H., U.S. Patent **3,458,420** (1969).
26. Slater, W. W., Thow, L. E., U.S. Patent **3,468,779** (1969).
27. Tawn, A. R. W., *Paint, Oil, Colour J.* (1969) 821.
28. Brockmann, F. J., Canadian Patent **863,924** (1971).
29. Munn, R. H. E., Holder, M. V., McNeeney, P., British Patent **1,235,975** (1971).
30. Bosso, J. F., Wismer, M., W. German Patent **2,003,123** (1971); **2,033,123** (1971).
31. McCoy, P. E., U.S. Patent **3,159,558** (1964).
32. Olson, D. A., *J. Paint Technol.* (1966) **38**, No. 499, 429.
33. Wismer, M., Bosso, J. F., *Chem. Eng.* (1971) **78**, 114.
34. Finn, S. R., Mell, C. C., *J. Oil Colour Chem. Assn.* (1964) 219.

RECEIVED May 28, 1971.

Theory: Introduction

GEORGE E. F. BREWER

Electrodeposition of macro-ions occurs on contact with the electrode of opposite polarity. During the deposition process many chemical and physical reactions may occur. Of these reactions, or as part of the mechanisms of deposition, the following should be considered by the paint formulator:

(1) Electrophoresis or migration of particles toward an electrode
(2) Electric discharge of macro-ions
(3) Coagulation of dispersion caused by pH change—*e.g.*, through anodically formed hydrogen ions near anodically forming deposits
(4) Oxidation on the anode
(5) Presence of coagulation-causing metal ions through electrically induced anode dissolution
(6) Organic electrode reactions such as Kolbe decarboxylation
(7) Reactions arising from the conversion of electric energy into heat at the electrode surface
(8) Dehydration of the colloid caused by electroosmosis.

While the organic macro-ions are required to form permanent films on the electrode which are not easily redispersible in the coating bath, the contact of the counterion with the electrode should not give rise to permanent deposition. Thus, alkali ions, amines, and ammonia are typical counterions for anodic deposition while acetate, oxalate, and hydroxyl ions are used for cathodic deposition processes.

As in all other electrodepositions, the film-formers will start to deposit at that section of the electrode which is nearest the electrode of opposite polarity. For instance, if a coating bath is housed in a cylindrical can and a test panel to be coated is submerged in the bath under completion of an electric direct current circuit, the deposition will start at the edges which are nearest the counter electrode (Figure 1). If the first formed electrodeposit offers higher electrical resistance than the original metallic substrate, then the electric current, seeking the path of least resistance, will cause electrodeposition on as yet uncoated areas until the entire surface of the substrate is coated.

The above described phenomenon is significant in two respects. First, it distinguishes the formation of electrodeposited coats by use of organic macro-ions from the electrodeposition of metal ions (plating). Secondly,

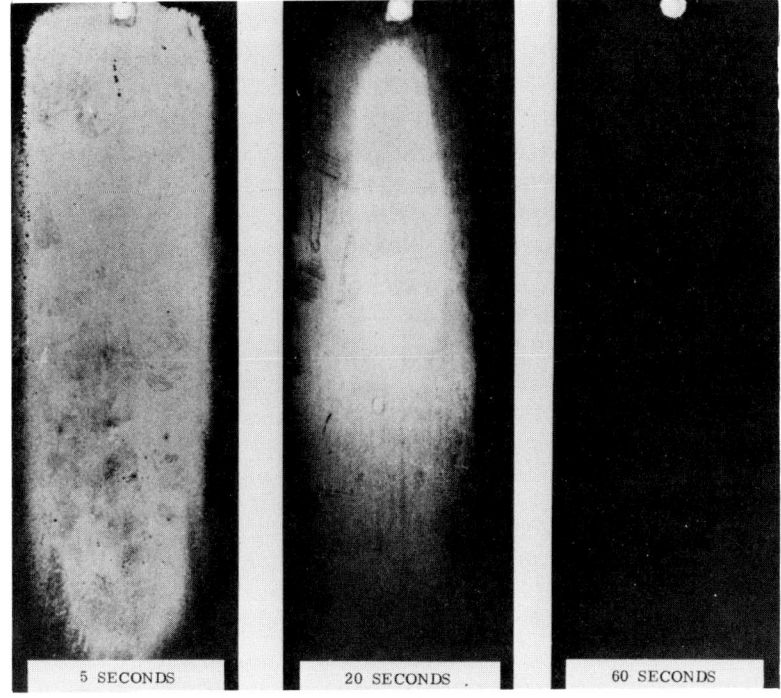

Figure 1. Electrodeposition of film-former

an important reason for the rapid, worldwide acceptance of the electrodeposition process is the built-in tendency to extend coats into extremely recessed areas of a workpiece. This desirable ability is usually called "throwing power."

Attention should, however, be given to the fact that the deposition conditions inside of a channel are different from the conditions on the outer surface of the same channel, not only regarding the electric field, but also regarding agitation, etc. It is, therefore, necessary to test the paint films deposited on recessed surfaces (1) in addition to testing films deposited on outer surfaces.

Literature Cited

1. Brewer, G. E. F., Strosberg, G. G., Horsch, M. E., "Prediction of Corrosion Protection from Salt Spray Tests on Throwing Power Strips," "Preprints," *Div. Org. Coatings Plastics Chem., ACS* (1967) **27** (1), 225–227; *J. Paint Technol.* (1967) **39** (512), 551–553.

9

Electrodeposition of Carboxyl Containing Copolymers

Basic Studies, Phase Growth, Counterion Fixation

A. E. RHEINECK[a] and A. M. USMANI

Polymers and Coatings Department, North Dakota State University, Fargo, N. D.

> *Two copolymers containing the potassium salts of acrylic or methacrylic acid were used in vehicles for electrodeposition at a constant voltage (125 volts). Zinc, steel, and tin were used as anodes. Up to 88.8% of the total charge transferred was used for anodic dissolution with the acrylic vehicle while 72.8% was used in anodic dissolution with the methacrylic vehicle. Electrodeposits contained only 0.48 to 0.77 wt % potassium; thus, counterion fixation was slight. NMR results indicate that electrodeposition has only a minor effect on tacticity and microstructure of the copolymers. Infrared results confirm that decarboxylation occurs especially at the higher voltages, but molecular weight determinations indicate that macroradical recombination is not extensive.*

Recently the electrodeposition of organic coating compositions has shown good potential, and the commercial exploitation as a method of coating application is now a reality. It has and will continue to make a firm impact in coating application operations such as the priming of automobile bodies, the coating of aluminum extrusions, and home appliance parts and related objects.

Historically, Crosse and Blackwell, Ltd. (*1*, *2*), developed and showed the possible potential of the applications of "oleoresinous lacquers" by an electric current. Today, commercial systems use various types of water soluble or water dispersible polymeric compositions. These are generally

[a] Deceased, August 10, 1971.

carboxyl bearing polymeric compositions inherently insoluble in water but capable of combining with organic or inorganic bases. The chemical compositions and properties of such systems are described by Gilchrist (3), Hagen (4), and Sullivan (5). When an electrical potential is impressed across their aqueous systems, a deposition on the anode will occur.

In addition to a solubilized or dispersed resin system, an electrodeposition bath may contain dispersed pigments, ionizable and other compounds as described by Gloyer et al. (6). Thus, ions and surface-charged particles in the broadest sense are involved. A successful continuous electrodeposition operation requires that all the components maintain a material balance. However, during electrodeposition the bath composition tends to change unevenly, and uneven changes may be critical. If the shielded areas on objects to be electrocoated are not properly coated, the coatings lack throwing power. Methods to determine and overcome these problems have been discussed by Brewer and co-workers (7–10) in their descriptions of the Ford Motor Co.'s electrocoating process. In a similar vein, an Italian process is described by Bono and Pagani (11).

The anodic electrode reactions have been studied by LeBras (12), Hagen et al. (13), and May and Smith (14). In a previous report (15), from our laboratory, electrode reactions occurring during electrodeposition in the case of a maleic adduct of a polymeric polyol-oleic acid ester neutralized with potassium hydroxide were described. The important role of metallic cations generated from the anode was elucidated and the extent of cation–polyion interaction was discussed.

Fresh electrodeposited coatings are cured by baking or may be cured by induction heating. Recently, two patents (16, 17) have been granted to the Ford Motor Co. These cover the curing by radiochemical methods and electrodeposited coating methods. This curing process merges electrodeposition and high energy curing. There is no doubt that this merger of processes represents a technological advance.

The objective of this paper is to elucidate the mechanism of electrodeposition at constant voltage when carboxyl containing acrylic copolymers are used as electrodeposition resins. Methods include determination of the mass of the electrodeposit as a function of time and charge transferred, determination of current variations with time, analysis of metal content of electrodeposits, and investigation of film structure by molecular weight determination and infrared and NMR spectroscopy. The metals analyzed include cations formed by anodic dissolution and potassium ions which are present as counterions in the electrodeposition bath. Since the retention of counterions in the electrodeposit may influence water and salt spray resistance of the film, potassium content is of particular interest.

Experimental

Electrodeposition. Two acrylic copolymers were used for the study; the compositions are indicated in Table I. The copolymerization was conducted at 125 ± 2°C with 2-butoxyethanol as the solvent. The catalyst was a 4% di-*tert*-butyl peroxide solution based on total monomer in both reactions. The final non-volatile content of both resin solutions was 70.0%. The acid values based on solids were 76.0 for copolymer A and 75.0 for copolymer M (as mg KOH/gram solids).

Table I. Monomer Mole Composition of Electrodeposition Resins

Monomers	*Copolymer A, Moles*	*Copolymer M, Moles*
Methyl methacrylate	1.5	1.5
n-Butyl acrylate	4.0	4.0
Acrylic acid	1.0	—
Methacrylic acid	—	1.0

Both resins were neutralized to the extent of 93% with potassium hydroxide and their solids adjusted to 7.0% with deionized water. The electrodeposition vehicle obtained from copolymer A will be henceforth referred to as EDV-A and that from copolymer M as EDV-M. The pH of both solutions was 8.3.

All depositions were run by the constant voltage technique. The voltage was applied prior to immersion of the anode into the electrodeposition cell. The anodes studied were tin, zinc, and cold rolled steel. All electrodes were 7.5 cm wide and immersed 7.5 linear cm. The total area of the immersed surface was 112.5 cm^2 per electrode. The deposition voltage was maintained at 125 volts, unless otherwise indicated, and the electrode residence time in the cells was varied from 15 to 90 seconds.

It was felt that the yields of the deposit per electrode on 112.5 cm^2 were too small for the analytical work. Therefore in each run, a duplicate was also prepared, and the deposited films combined. In other words, the total deposition surface was $2 \times 112.5 = 225$ cm^2 for the results to be described.

All panels after electrodeposition were washed and dried in a vacuum oven at 110 ± 2°C. The quantity of electricity in microfarads, μF, was determined from the coulombic curves. For two similar runs, they were added to get the total μF. Likewise, the weights of two similar deposits were determined and combined to get the total weight.

Recovery and Analysis of Electrodeposits. The electrodes with the deposited films after a vacuum drying were exhaustively extracted with methyl ethyl ketone in a Soxhlet extractor. The extraction was continued until no organic material remained on the electrodes. This was checked by recording a surface spectrum of the electrode in the infrared region (*18*). Methyl ethyl ketone was evaporated from the solubilized polymer films. The solids were then ashed at about 600°C in a muffle furnace for six hours. A small portion of perchloric acid was used to assist the removal of the organic matter. As controls, polymers A and M *per se*, did not leave any residue when similarly ashed. The inorganic residue from ashed electrodeposited films was dissolved in dilute nitric acid.

Metal contents of the ashed solutions, thus prepared, were determined by atomic absorption spectroscopy (19). For atomic absorption analysis, the liquid solution sample was atomized in a flame. Metallic ions present were reduced to neutral atoms. An Osram lamp was used to obtain potassium spectrum while hollow cathode lamps of iron and zinc were used to obtain the spectra of iron and zinc, respectively (19). Calibration curves were prepared for all the metals with analytical grade inorganic salts. The wavelengths used for potassium, zinc, and iron were 4665, 2139, and 2483 A, respectively.

A Perkin-Elmer model 421 spectrophotometer was used to record the infrared spectra of electrodeposits. These spectra were analyzed to follow the change in carbonyl content with deposition voltage. The standard base line technique coupled with rationing the peak-areas was used. The method is described by Rheineck et al. (15, 18). The carbonyl content Y_{CO} is expressed as follows:

$$Y_{CO} = \frac{\text{Area of carbonyl peak}}{\text{Area of hydrocarbon peak}} \times 100$$

NMR spectra were obtained on a Varian A 60A high resolution spectrometer. The polymer concentration was 20% in deuterated chloroform. Tetramethylsilane was used as an internal standard, and spectra were recorded at ambient temperature.

Number average molecular weights, M_n, were determined with a Mechrolab vapor phase osmometer. Benzil, recrystallized three times from methanol was used to obtain the calibration curve.

Results and Discussion

Mass of Electrodeposit. Electrodeposition involves phase extension (including phase dissolution). Cohn (20) studied the tarnishing of silver metal by halogens and by applying the disorder theory obtained two time dependencies. Mathematically, this may be represented as follows:

$$am + m^2 = p \cdot t \qquad (1)$$

where: a is constant

The above concept can be used in electrodeposition which is essentially a phase growth process. Two limiting cases are possible:

m is mass of phase growth

p is proportionality constant

t is time of reaction

Case 1: when t is small; $m^2 \ll am$. Thus, Equation 1 reduces to:

$$m = \frac{p}{a} \cdot t \qquad (2)$$

Case 2: when t is large; $m^2 \gg am$. Then, Equation 1 becomes:

$$m = (p \cdot t)^{1/2} \qquad (3)$$

In Figures 1 and 2, mass deposited *vs.* time and square root of time are plotted for zinc anode and EDV-A and EDV-M respectively. Both the above limiting cases (Equations 2 and 3) are satisfied. Olson (*21*) has suggested an unimpeded phase growth during early stages of deposition. During advanced deposition stages diffusion through the growing phase becomes important. It is also to be noted from these graphs that the square root time dependency is valid up to 45 sec of deposition; the intercept changes, and then the relationship holds again. This invalidates the conceptual constancy of p. However, our results suggest that p might have two values, X and Y, which are time and process dependent. In the case of a cold rolled steel anode, the above described dependencies were not obeyed for both electrodeposition vehicles.

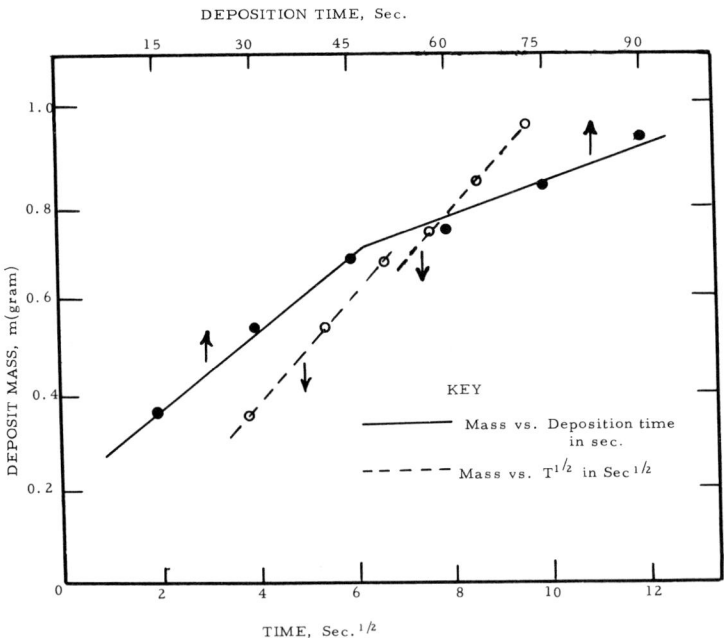

Figure 1. Deposit mass vs. time and square root of time for zinc anode and EDV-A

Finn and Hansip (*22*) have shown mathematically that the reciprocal of square of current should give a linear curve when plotted against deposition time. This relationship is only valid if the specific resistance of the deposit remains constant during the entire deposition period. Figures 3 and 4 are plots of reciprocal of the square of current, $1/c^2$ vs. deposition time, t, at a deposition voltage of 125 volts for zinc and cold rolled steel electrodes respectively. The above relationship is followed reason-

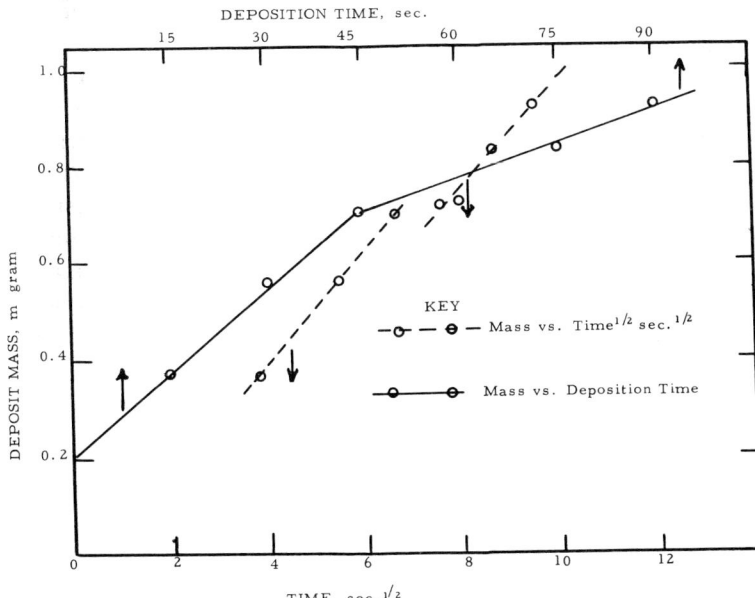

Figure 2. Deposit mass vs. time and square root of time for zinc anode and EDV-M

ably well in the case of cold rolled steel and therefore, the specific resistance of the deposit appears to remain constant throughout the deposition period. This suggests that the film does not necessarily become compacted because of electroosmosis. However, for the zinc electrode, a departure from linearity is observed. This suggests that the film compacts during deposition by electroosmosis.

The dependence of deposition mass on charge transferred has been previously discussed in detail by us (15). The overall reaction in electrodeposition involves electrochemical reactions in conjunction with heterogeneously catalyzed reactions. The rate laws applicable in electrodeposition were also discussed. For the total electrodeposition process, the following equations were presented:

$$m = k \cdot \mu F \tag{4}$$

and its differential form:

$$\frac{dm}{d\mu F} = k \tag{5}$$

If a plot of m, the mass of polymer deposited *vs.* charge transferred in microfarads is a straight line, then Equations 4 and 5 are valid for the deposition. The slope, k, will be the overall deposition constant with units $g\mu F^{-1}$. The plot of mass deposited, m, vs. charge transferred in μF for EDV-A and EDV-M were generated for all the three electrode materials (graphs not included here; for details *see* Ref. *15*). Straight lines were obtained with zinc and tin electrodes and rate of deposition calculated. For some unknown reasons, the m vs. μF plot for cold rolled steel did give some scattering, and hence the rate constant, k_{CRS} is not reported. Rate constants, k_{Zn} and k_{Sn} for EDV-A were 8.0×10^{-4} and 7.42×10^{-4} $g\mu F^{-1}$ respectively. The same rate constants were obtained for EDV-M— *viz.*, k_{Zn} equals 8.0×10^{-4} and k_{Sn} equals 7.42×10^{-4} $g\mu F^{-1}$.

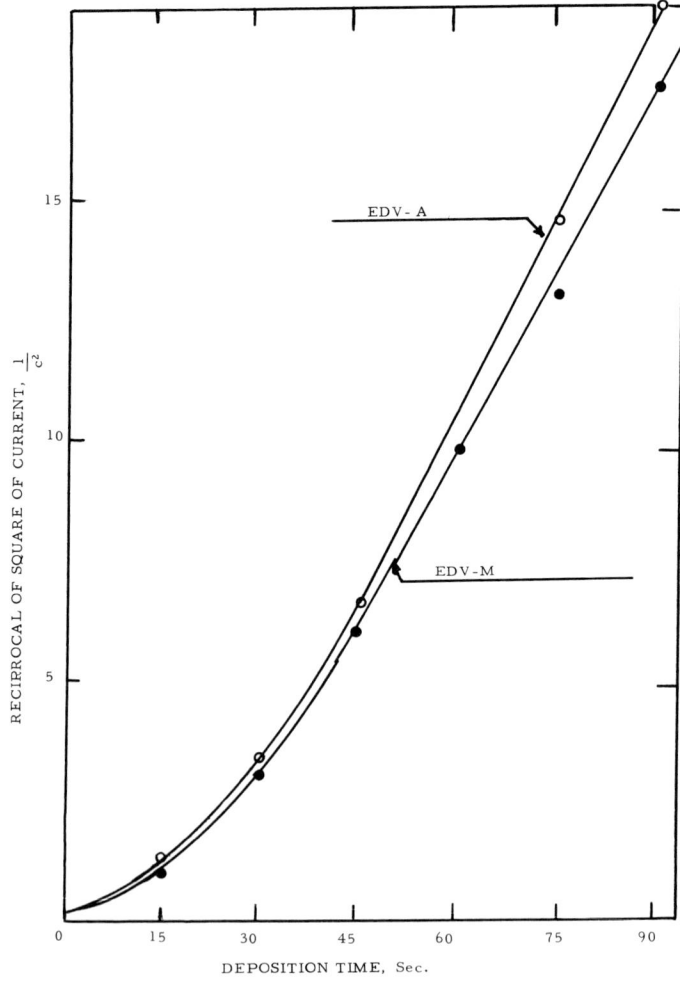

Figure 3. $1/c^2$ vs. deposition time for zinc anode

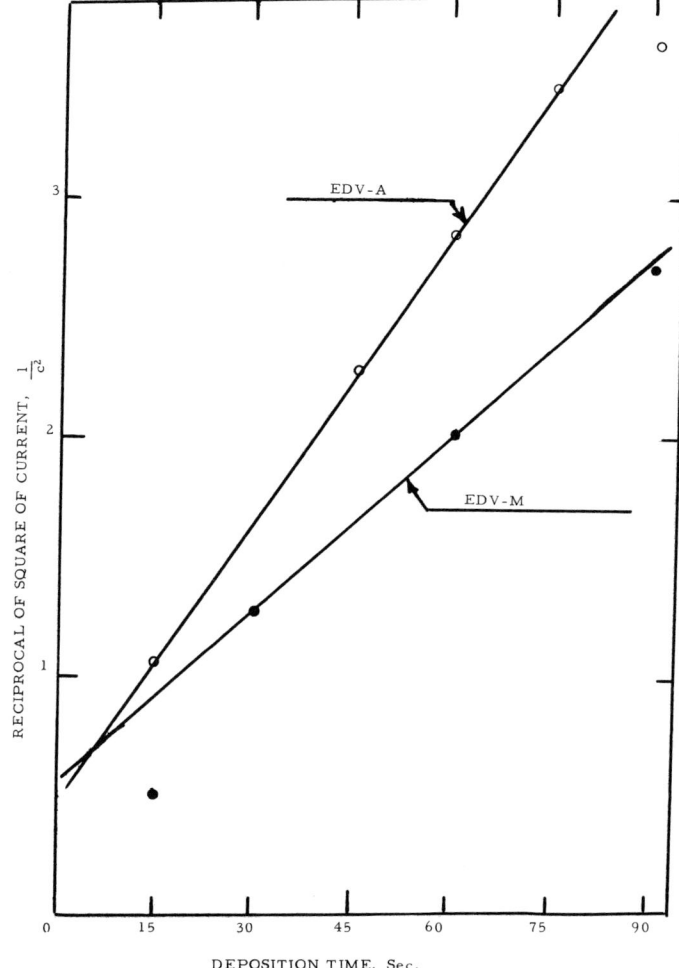

Figure 4. $1/c^2$ vs. deposition time for cold rolled steel anode

Cation–Polyion Interaction. Charge transfer in electrodeposition may involve anodic dissolution, oxidation of polymeric polyions, and redox reaction(s) which may involve other anions present in the cell. A part of the charge involved in electrodeposition cell may be non-Faradic such as a charging of the double layer. A plot of zinc content in the polymer deposit vs. charge transferred gives a linear relationship as shown in Figure 5. The slope of the lines for EDV-A and EDV-M are 29.0 and 23.8 grams/Faraday respectively. The slope of Figure 5 is a measure of charge used for the dissolution of the anode. The percent of charge transfer used for the anodic dissolution of zinc can be calculated from

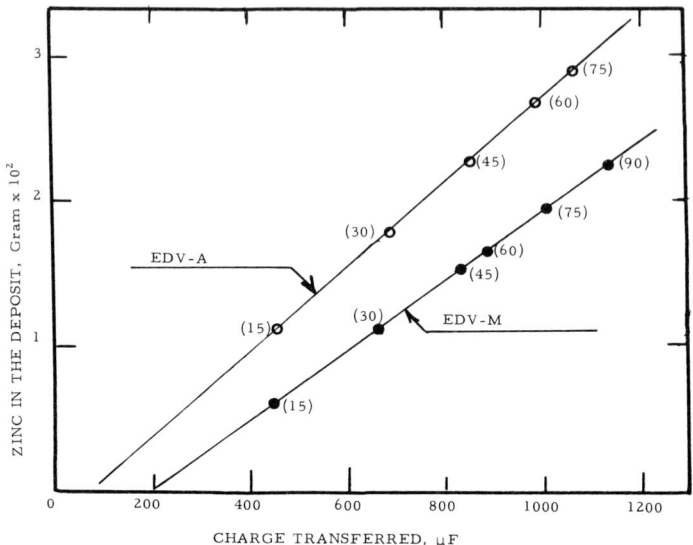

Figure 5. Anodic dissolution of zinc. Deposition time in seconds inscribed.

the slope and the equivalent weight of zinc. It is 88.8% and 72.8% for EDV-A and EDV-M respectively. This difference can be explained on the basis of more stable macroradicals formed by the oxidation of polyions and followed by more decarboxylation in EDV-M than EDV-A.

The plots in Figure 5 should theoretically pass through the origin. There seems to be a delay of 90 and 200 μF for the generation of Zn^{2+} in EDV-A and EDV-M respectively. It cannot be explained on the basis of easily oxidizable contaminants in the cell since none were present. Neither can the delay be attributed to non-Faradic charge since a straight line was generated in the rate plot. During this delay period, the current density is at an all time high, and the oxidation of the ion:

$$R-C{\overset{O}{\underset{O}{\diagdown}}} \quad \text{(where R is polyion moiety)}$$

does occur. A larger delay in EDV-M is caused by the favored stability of the macroradical derived therefrom. This does substantiate the explanation offered earlier in connection with the difference in the slope. Thus, EDV-M is more susceptible to anodic oxidation than EDV-A. This would seem reasonable from structural considerations since a 3° macroradical (produced from EDV-M) would be more stable than a 2° macroradical (produced from EDV-A). The macroradicals produced as a result of oxidation and decarboxylation of EDV under consideration does disproportionate to produce decarboxylated polymer deposits as evidenced by molecular weight studies later discussed.

Figure 6 shows the iron content of the polymer deposit *vs.* charge transferred for EDV-A. A straight line was obtained, but departure from linearity occurred at high deposition times. The slope of the straight line for this plot is 1.27 gF^{-1} and a delay of approximately 50 μF. From the slope of the line and the equivalent weight of iron, the percent charge used up towards dissolution of iron to produce cations was calculated. This percent was found to be 6.38. When the deposition time is high, the overall current density is low. This results in more anodic dissolution of iron (*13, 15*).

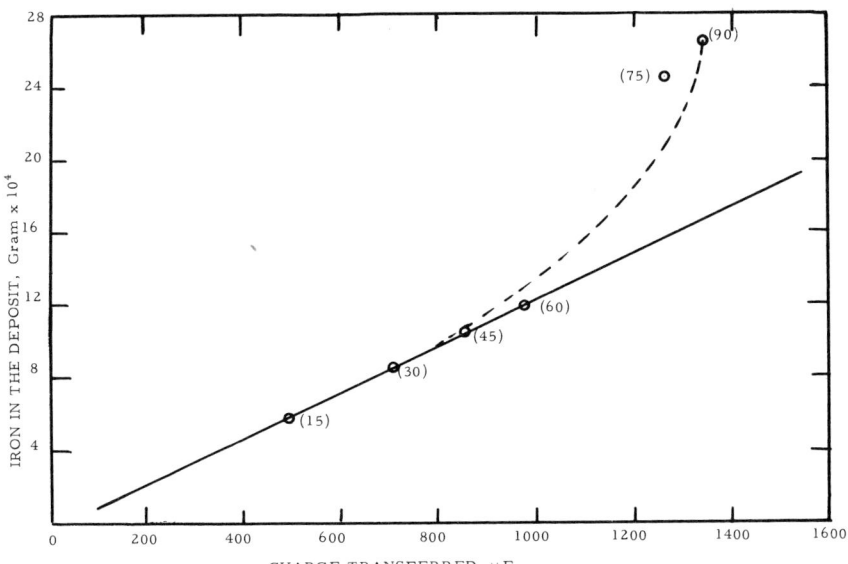

Figure 6. Anodic dissolution of cold rolled steel in EDV-A. Deposition time in seconds inscribed.

The percent reaction arising from Zn^{2+}–polyion interaction was calculated (*15*) from stoichiometry and was in the range of 39–40% for six deposits of EDV-A. The striking fact is that percent interaction is more or less independent of deposition time or the current density. For EDV-M, the percent interaction was lower, namely, in the range of 23.0 to 27.0% (increasing with time of deposition). Table II summarizes percent interaction for cold rolled steel anode with EDV-A. The percent interaction remains constant up to the straight-line portion of Figure 6, and then it suddenly increases as expected.

Table II. Extent of Cation–Polyion Interaction as a Function of Deposition Time for Cold Rolled Steel Anode

Deposition Time, sec	% Reaction caused by Cation–Polyion Interaction
15	3.19
30	3.20
45	3.35
60	3.27
75	4.75
90	7.50

Counterion Fixation in Electrodeposited Polymer. There are sketchy reports in the literature about small amounts of the counterion retained in the deposited polymer. Tawn and Berry (23) have reported triethylamine content in deposited films at various stages during deposition. They have shown that percent triethylamine retained in the deposited films falls off dramatically with the passage of the current. In their study, the constant current technique was used, and the current density was quite low. Presently used U.S. electrodeposition processes employ high voltages. The bound counterions are forced out of the coagulated resin by electroosmosis (24).

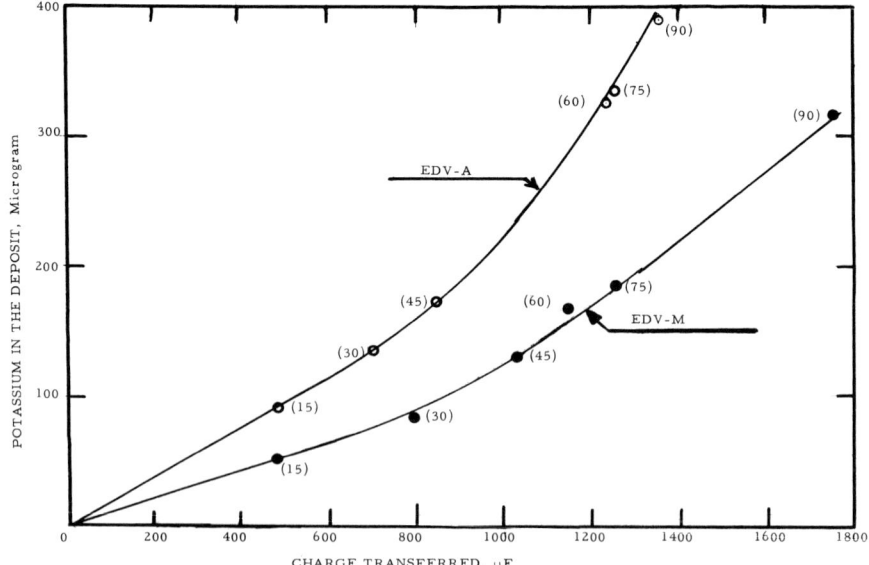

Figure 7. Counterion fixation in electrodeposits on cold rolled steel anode. Deposition time in seconds inscribed.

In view of this, a quantitative study of the counterion fixation in the deposited polymers was in order. In Figure 7, the plot of potassium counterions retained in the deposit vs. charge transferred is shown. The retention of counterions increases with increase in the deposition time. Thus, a lower average current density appears to be conducive to the retention of counterions in the deposit.

Percent potassium counterion fixation in electrodeposits is defined and expressed as follows:

$$\% \text{ K fixation} = \frac{\text{gram K in the deposit}}{\text{(total) gram deposit}} \times 100$$

The percent of potassium contained originally in EDV-A can be computed from the acid value of polymer A (76.0 mg KOH/gram polymer) and the extent of neutralization (93%). It was calculated that EDV-A on a solids basis, contains 4.65% of potassium. The percent K fixation defined as above is known for deposits of EDV-A. Thus, the percent absolute potassium retention, based on electrodeposition vehicle, can be calculated as follows:

$$\% \text{ absolute K retention} = \frac{\% \text{ K fixation}}{\% \text{ K in EDV-A}} \times 100$$

Results are summarized in Table III for such calculations for EDV-A.

Table III. Percent Potassium Fixation in Deposits and Percent Absolute Potassium Retention for EDV-A at 125 Volts

Deposition Time, sec	% K Fixation × 10^2	% Absolute K Retention
15	2.40	0.518
30	2.36	0.508
45	2.28	0.490
60	2.22	0.480
75	3.10	0.668
90	3.54	0.762

There are three possible mechanisms for the fixation of counterions: (1) Formation of intimate ion pairs with

(2) Partial retention of loose ion pair with the following:

$$R-C\begin{matrix} \diagup O \\ \diagdown O \end{matrix} \Bigg\} \quad - \text{ or,}$$

(3) Trapping of counterions in the collapsed macromolecule chains.

The low levels of potassium fixation suggest that electroneutrality within the deposit must be attained by some positive species other than K^+. If all unreacted carboxylate anions were associated with potassium ions as suggested by mechanisms (1) and (2) above, the potassium fixation would be higher than that observed. The low levels of potassium retention observed in this study would suggest that counterion fixation may have only a minor influence on water and salt spray resistance of electrodeposited films.

Electrode Reactions and Mechanism of Electrodeposition. Finn and Mell (25) describe the origin of charge in various electrodeposition systems and propose an ionic mechanism for the deposition. In a previous paper (26), we proved by infrared spectroscopy that an ionic mechanism is indeed operative in carboxyl containing systems. We observed the changes in the absorption spectra between 1600 and 1525 cm^{-1}. These may be caused by the asymmetric stretching of

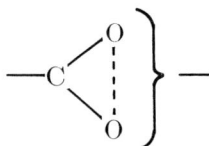

(27). In films of EDV's laid by flow out, a strong absorption was observed at 1560 cm^{-1} whereas this absorption was not present in spectra of electrodeposited films. This indicates that an ionic mechanism involving carboxylate anions is operative in electrodeposition. The term ionic mechanism is used in a broad sense. However, this does not exclude any reactions involving free radicals derived from the carboxylated polyions. Thus, anions and radicals derived therefrom are involved in the mechanism. To avoid confusion, we therefore, propose to name it as anionic-free radical mechanism (AFRM).

An impression of dc voltage on EDV-A or EDV-M solutions will result in the migration of ions. The carboxylated polymer anions will move and concentrate at or near the anode. Potassium ions will migrate to the cathode. When the carboxylated polymer anions are near the anode, the following reactions (15) occur:

(1) Protonation of polyanions:

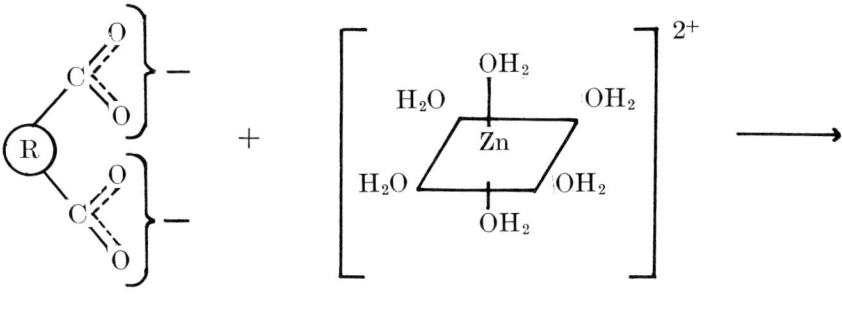

Where ⓡ is the polyanion moiety

Protonation near the anode will occur to a greater extent than in the bulk as a result of the local pH decrease associated with

$$4\ OH^- \rightarrow 2\ H_2O + O_2$$

Protonation results in precipitation, and the proton capture product contributes substantially to the total deposit.

(2) The interaction of polyanions with multivalent cations generated by anodic dissolution is represented, for example, in the case of zinc anode as follows:

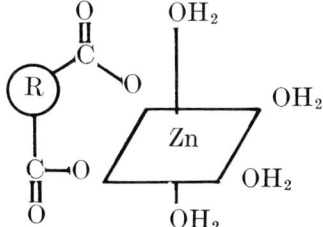

(3) Anodic oxidation of polyanions may yield carboxyl radicals. These may undergo decarboxylation to macroradicals. The macroradicals may combine or disproportionate to give the deposit.

Radical Formation

Coupling Reaction (combination)

Disproportionation

(disproportionation; H* shift from one macroradical to another macroradical).

(4) Intramolecular reactions leading to formation of γ-lactone as shown on p. 145, top.

Molecular Weight Studies. It was not possible to determine number average molecular weight, $\overline{M}_n$, of deposits on the zinc anode. There was no problem in completely removing a deposited polymer from the zinc anode by exhaustive extraction with methyl ethyl ketone. When the extracted solution was cooled to room temperature, precipitation invariably occurred.

9. RHEINECK AND USMANI *Carboxyl Containing Copolymers*

[Disproportionation reaction scheme showing polymer chain with CH$_3$ and COOH groups]

[Cyclization reaction scheme showing polymer chain with CH$_3$ and COOH groups forming C=C]

[Resulting cyclic structure with ester linkage]

The deposits on cold rolled steel anode were soluble in methyl ethyl ketone. In Table IV, $\overline{M}_n$ of the deposits are compiled. Thus, combination of macroradicals can be excluded in favor of the disproportionation as evidenced by a fair constancy of $\overline{M}_n$.

Table IV. Number Average Molecular Weight, $\overline{M}_n$ of Electrodeposits on Cold Rolled Steel Anode

No.	System	Deposition Time, sec	$\overline{M}_n \times 10^{-3}$
1	Polymer A	—	2.136
2	EDV-A	15	2.318
3	EDV-A	30	2.133
4	EDV-A	45	2.561
5	EDV-A	60	2.067
6	EDV-A	75	2.130
7	EDV-A	90	1.827
8	Polymer M	—	1.997
9	EDV-M	15	1.930
10	EDV-M	30	1.930
11	EDV-M	45	1.915
12	EDV-M	60	2.016
13	EDV-M	75	1.965
14	EDV-M	90	2.000

NMR and Infrared Studies. Deposits on cold rolled steel and zinc anodes were soluble and insoluble in deuterated chloroform, respectively. The former polymer solutions contained some combined iron which will interfere with the NMR signals. The NMR spectra of deposits on a cold rolled steel anode were not good because the base line was shifted in the downfield region. The main purpose of this work was to determine changes in tacticity and microstructure caused by deposition. The chemical shifts and the spectra of the polymer and electrodeposits derived therefrom were very similar (28). Thus, the tacticity and the microstructure of a polymer are not influenced by electrodeposition.

Although a deposition voltage of 125 volts was used in all work described above, the voltage was varied in one series of experiments to determine its effect on the extent of decarboxylation. In Figure 8 the carbonyl content, Y_{CO}, is plotted against deposition voltage for both EDV's. The data presented here are for tin anodes and a deposition time of 90 seconds. The use of infrared spectroscopy for following decarboxylation during electrodeposition is discussed at greater length elsewhere (15).

As pointed out, the formation of macroradicals is favored at high deposition voltages. These macroradicals disproportionate, and the disproportionated polymer may cyclize if the β-carbon has a pendant carboxyl group.

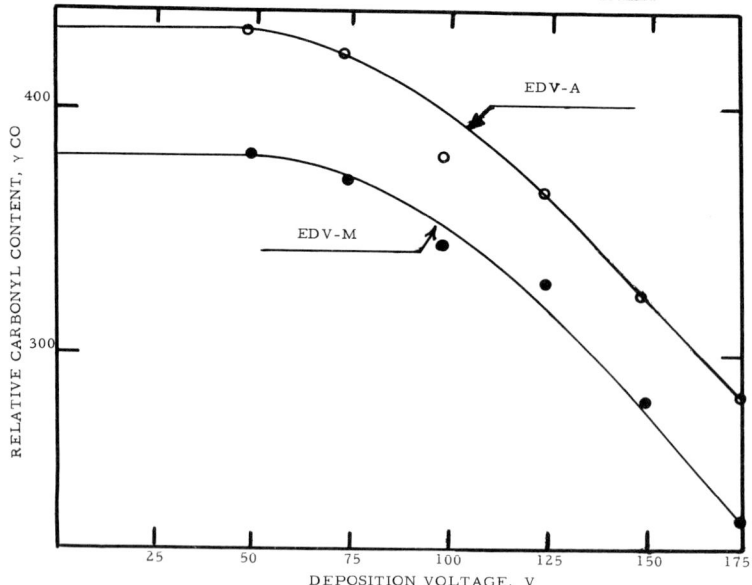

Figure 8. Relative carbonyl content, Y_{CO}, vs. deposition voltage for tin anode

No adsorption for γ-lactone was observed in electrodeposits of EDV-A and EDV-M, and hence γ-lactones are not formed. Previously (15), we reported formation of a β-γ unsaturated acid and subsequent cyclization to γ-lactone in the deposition of maleic adduct of a polymeric polyol-oleic acid ester neutralized with potassium hydroxide.

Summary

Mass of the electrodeposit was linearly related to the square root of time for the zinc anode but not with cold rolled steel anode. A current–time relationship study indicates that the deposit on zinc anode becomes compacted by electroosmosis. Deposits on cold rolled steel anode are less prone to compact.

The rate of deposition and the extent of anodic dissolution were studied. In the case of zinc anode, 88.8% of charge involved is used for the anodic dissolution with EDV-A (vehicle based on acrylic copolymer containing acrylic acid). It is 72.8% for EDV-M (vehicle based on acrylic copolymer containing methacrylic acid). A delay of 90 and 200 μF was established by extrapolation before the generation of Zn^{2+} starts in EDV-A and EDV-M, respectively. In a similar vein, the anodic dissolution of cold rolled steel was studied. Percent reaction caused by cation–polyion interaction was more or less independent of deposition time up to 60 sec and on the order of 3.2–3.3%. For 75 and 90 sec, anomalous percents were obtained—4.75 and 7.5, respectively.

Counterion fixation was quantitatively studied. Percent potassium fixation is expressed as grams potassium in the deposit per gram of the deposit. Fixation was very small, on the order of 10^{-2}%. Percent absolute potassium retention is expressed as percent potassium fixation divided by percent potassium originally present in the EDV. It was in the range 0.48–0.77%. Possible mechanisms for fixation of counterions are discussed. Trapping of counterions in the collapsed macromolecule seems to be the primary reason.

Anions and free radicals are involved in the mechanism of electrodeposition. Therefore, electrodeposition involves an anionic-free radical mechanism, (AFRM).

Infrared spectroscopy was used to study film structure as a function of deposition voltage. NMR spectroscopy was used to check if a change in tacticity or microstructure occurs when the polymer in a suitable form is subjected to electrodeposition. No such structural or stereoregulating changes were observed.

Acknowledgment

Financial assistance by Control Data is gratefully appreciated.

Literature Cited

1. Crosse and Blackwell, Ltd., British Patent **455,810** (1936).
2. Crosse and Blackwell, Ltd., British Patent **496,945** (1937).
3. Gilchrist, A. E., U. S. Patent **3,230,162**.
4. Hagen, J. W., *J. Paint Technol.* (1966) **38**, 436.
5. Sullivan, M. R., *J. Paint Technol.* (1966) **38**, 424.
6. Gloyer, S. W., Hart, D. P., Cutforth, R. E., *J. Paint Technol.* (1965) **37**, 113.
7. Brewer, G. E. F., Strosberg, G. G., Madejczyk, L. A., Hines, R. F., *J. Paint Technol.* (1966) **38**, 499.
8. Brewer, G. E. F., Horsch, M. A., Madrarasz, M. F., *J. Paint Technol.* (1966) **38**, 452.
9. Brewer, G. E. F., Hamilton, C. C., Horsch, M. E., *J. Paint Technol.* (1969) **41**, 114.
10. Burnside, G. L., Brewer, G. E. F., Strosberg, G. S., *J. Paint Technol.* (1969) **41**, 431.
11. Bono, M., Pagani, D., *J. Paint Technol.* (1967) **40**, 123.
12. LeBras, L. R., *J. Paint Technol.* (1966) **38**, 85.
13. Hagen, J. W., Orttung, F. N., Chow, S. E., *Paint Varnish Prod.* (1967) **67**, 48.
14. May, C. A., Smith, G., *J. Paint Technol.* (1968) **40**, 493.
15. Rheineck, A. E., Usmani, A. M., *J. Paint Technol.* (1969) **41**, 597.
16. Turner, A. H., U. S. Patent **3,501,390**.
17. Smith, A. G., U. S. Patent **3,501,391**.
18. Rheineck, A. E., Peterson, R. H., Sastry, G. M., *J. Paint Technol.* (1967) **39**, 484.
19. Walsh, A., *Spectrochim. Acta* (1965) **7**, 108.
20. Cohn, G., *Chem. Rev.* (1948) **42**, 527.
21. Olson, D. A., *J. Paint Technol.* (1966) **38**, 429.
22. Finn, S. R., Hansip, J. A., *J. Oil Colour Chem. Assoc.* (1965) **48**, 1121.
23. Tawn, A. R. H., Berry, J. R., *J. Oil Colour Chem. Assoc.* (1965) **48**, 790.
24. Mysels, K. J., "Introduction to Colloid Chemistry," Interscience, New York, 1959.
25. Finn, S. R., Mell, C. C., *J. Oil Colour Chemists Assoc.* (1964) **47**, 219.
26. Rheineck, A. E., Usmani, A. M., "Preprints," ACS Division of Organic Coatings and Plastics Chemistry (1967) **27**, 193.
27. Nakanishi, K., "Infrared Absorption Spectroscopy," Holden-Day, New York, 1964.
28. "Polymer Handbook," J. Brandrup and E. H. Immergut, Eds., section V-5, Interscience, New York, 1967.

RECEIVED May 28, 1971.

10

Electrochemistry of Polymer Deposition

ZLATA KOVAC-KALKO

PPG Industries, Inc., Coatings & Resins Division, Springdale, Pa. 15144

> *The electrodeposition of oil-modified polyesters and epoxy esters was investigated at constant current density, j, and at constant preset applied voltage, E_{app}. At constant j, the electrodeposition starts after an induction time, τ. Film thickness and electrode potential increase linearly with time, t, for $t > \tau$. Coulombic efficiency and film resistivity are independent of t. Coulombic efficiency increases and film resistivity decreases with increased j. At constant E_{app}, at the beginning of electrodeposition the electrical field is high, and the growth of the film follows the logarithmic time law. The film resistance is non-ohmic. With increased thickness the electrical field decreases, the growth follows the $\sqrt{t}$ law, and resistance becomes ohmic. Coulombic efficiency is independent of t but increases with increased E_{app}.*

Since the classical work of Fink and Feinleib (1) many publications have appeared on the electrochemical aspects of this process (2-8). Our interest was to find out how the electrodeposition of polymers begins, what law or laws determine the growth of films, and how anode potential, coulombic efficiency, and film resistance depend upon plating time and voltage used in commercial practice. In addition to constant voltage experiments characteristic of industrial use, constant current measurements were also made to obtain additional information.

Experimental

The electrodeposition of oil-modified polyesters and epoxy esters has been investigated at constant applied voltages, E_{app}, and at constant current densities, j.

Experiments were carried out in a stirred emulsion at constant temperature. Anode potentials, E_a, were measured with respect to a reference electrode (saturated calomel or Pt electrode) via a Luggin capillary (to avoid the iR-voltage drop through paint) using a Keithley 660 electrometer. Currents were measured with a Keithley 600B electrometer. Both anode potential and current were recorded on a dual channel Brush recorded (Mark 280) as a function of electrodeposition time. A typical j–t and E_a–t graph is shown for the beginning of electrodeposition in Figure 1. The amount of charge flow was recorded with a coulometer (Vari-Tech model VT-1176B).

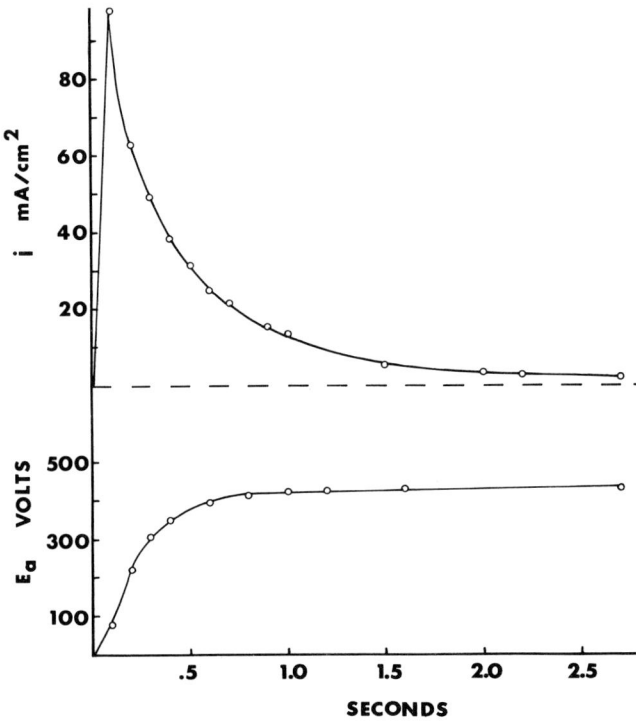

Figure 1. Current–time and anode potential–time curves for oil-modified polyester systems at $E_{app} = 450$ volts

Substrates were prebaked and weighed prior to electrodeposition. Plating areas were either 4 or 115 cm². A stainless steel cathode was placed parallel to the anode. The thickness of deposited films after baking were determined using a Permascope (Twin City Testing Co.). Also the weight of baked films was measured.

Results and Discussion

Beginning of Electrodeposition. Electrodeposition can be carried out at a constant current density (j) or at a constant voltage. The current densities generally used in electrodeposition are on the order of a few ma/cm². When the current is applied, the electrochemical reaction starts. We will assume that the main reactions are:

Anodic:
oxidation of water

$$2 H_2O = 4 H^+ + O_2 + 4e^- \quad (1)$$

and dissolution of substrate

$$M = M^{n+} + ne^- \quad (2)$$

Cathodic:
discharge of water

$$2 H_2O + 2e^- = 2OH^- + H_2 \quad (3)$$

The concentration of these ions at the interface is given by Sand's equation:

$$C_{x=0} = C_b + \frac{2j}{nF}\sqrt{\frac{t}{\pi D}} \quad (4)$$

where $C_{x=0}$ is the concentration at the interface, C_b is the bulk concentration, j is the current density, t is time, F is Faraday's constant, and D is the diffusion coefficient. The concentration of H^+ and metallic ions at the anode and OH^- ions at the cathode will increase with time and current density. The production of these ions will give rise to a concentration gradient across a boundary layer adjacent to the electrode. Diffusion processes set in to diminish this increase in concentration—some ions diffuse away. For anionic deposition, after a certain time, τ, known as the induction time, concentration of H^+ ions will be high enough to reach the solubility product, for a given system—i.e.,

$$[H^+](RCOO^-) = Ka \quad (5)$$

After this critical concentration is reached, the film will form at the anode. This is indicated by an increase in E_a with t (Figure 2).

From Equation 4 it follows that $j\sqrt{\tau}$ is a constant for a given system since n, F, π, and D are constants. So $j\sqrt{\tau}$ is a characteristic for a given substrate and polymer emulsion (5, 6). From Figure 3 it can be seen that $j\sqrt{\tau}$ is smaller (3.0×10^{-3} amps sec$^{1/2}$ cm^{-2}) on untreated steel than on Zn-phosphated steel (4.9×10^{-3} amp sec$^{1/2}$ cm^{-2}). Dissolution of the substrate can account for these differences because of the higher charge (Fe^{2+} or Fe^{3+} vs. H^+) which is more effective in coagulation.

From Sand's equation it is possible to calculate the interfacial concentration of H^+ ions and the pH at the electrode surface with Zn-phosphated steel. The pH at the surface was calculated to be 2.2 (pH bath = 8.9).

In contrast to $j =$ constant, experiments where it takes a few seconds for the formation of polymer film to start at a constant voltage, E_{app}, an electrode is completely covered with a film within a fraction of a second. This is caused by the large currents flowing at the beginning of electrodeposition. (The peak currents are on the order of 10–100 ma/cm² depending upon voltage applied and the conductivity of the paint emulsion, which are between 300–3000 mhos.)

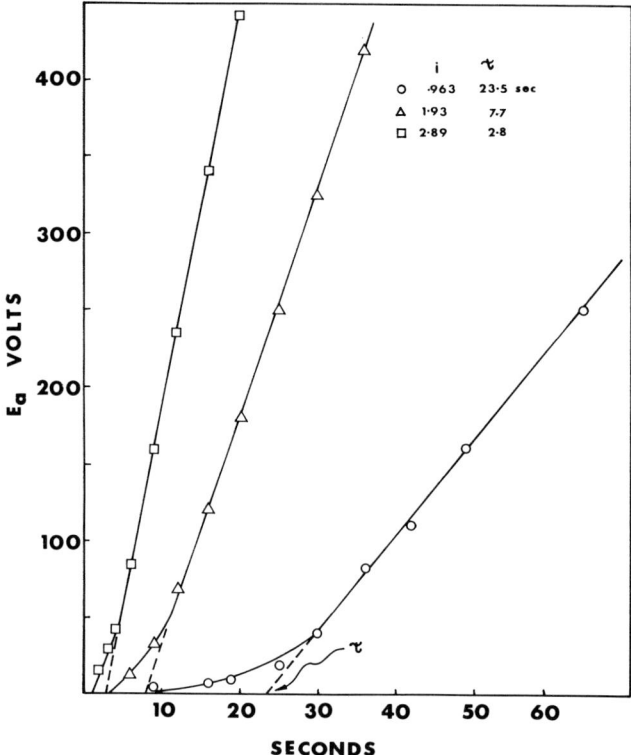

Figure 2. *Anode potential vs. deposition time in oil-modified polyesters at constant current density. Zinc-phosphated steel substrate.*

Growth of Film. If the film is an electronic insulator, it cannot transport the electrons which are required for Reactions 1 and 2. Therefore, the charge transfer can take place only at the metal/film interface. The ions formed in Reactions 1 and 2 then carry current through the film. They react chemically with the carboxylic ions arriving from the bath, giving rise to the formation of new layers of film. Hence, film thickness increases.

The transport of the ions through the film is caused by the presence of a high electrical field. One may ask, what is the most general relation between ionic flux or current density and the electrical field in any ionic

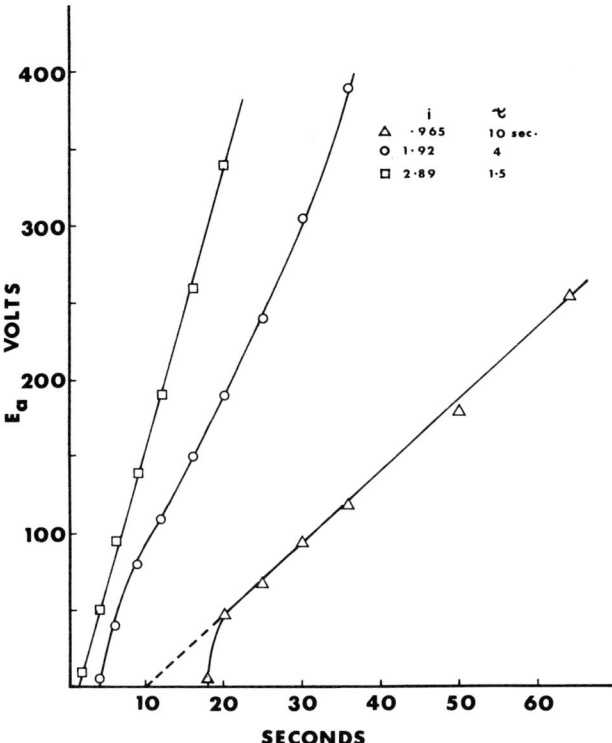

Figure 3. Anode potential vs. deposition time in oil-modified polyesters. Untreated steel substrate.

conductor? The answer according to the textbooks of electrochemistry is (9):

$$j = A \sinh \frac{qa\vec{F}}{kT} \tag{6}$$

where j is current density, A is a constant given by the conductivity of a system, q is the charge on an ion, a is the distance traveled by an ion between successful jumps, $\vec{F}$ is the electrical field, k is the Boltzmann constant, and T is absolute temperature. The product kT is the measure of thermal energy.

Equation 6 can be reduced to the simpler forms in the following special cases (10, 11):
(1) Low field approximation, when

$$\frac{qa\vec{F}}{kT} << 1$$

i.e., when the work done by electrical field on an ion is much smaller than thermal energy. In this case

$$\sinh \frac{qa\vec{F}}{kT} = \frac{qa\vec{F}}{kT}$$

and

$$j = \frac{1}{\rho} \vec{F} \tag{7}$$

i.e., current density is a linear function of the field. This is Ohm's law; where ρ is the specific resistivity of the wet film. Electrical field is defined as E_a/δ (*11*) where E_a is anode potential and δ is thickness of wet film.

The rate of increase of film thickness is:

$$\frac{d\delta}{dt} = j \frac{M}{nFd} \tag{8}$$

where M/nF is the electrochemical equivalent weight and d is density of a film.

At $j =$ constant,

$$d\delta/dt = \text{a constant or } \delta = j \frac{M}{nFd} (t - \tau) \tag{9}$$

Film thickness after induction time τ, increases linearly with time, as does E_a (*cf.* Figures 2, 3).

At $E_{app} =$ constant,

$$\frac{d\delta}{dt} = \frac{\text{constant}}{\delta}$$

or after integration

$$\delta = \text{constant} \sqrt{t} \tag{10}$$

i.e., thickness increases linearly with square root of time.
(2) High field approximation (*10*):

$$\frac{qa\vec{F}}{kT} \gg 1$$

i.e., electrical field is much greater than thermal energy; then

$$j = \frac{A}{2} \exp \frac{qa\vec{F}}{kT} \tag{11}$$

i.e., current is an exponential function of the field—non-ohmic behavior. In this case at $E_{\mathrm{app}} = $ constant (*10*):

$$\delta = \text{constant} \ln t \tag{12}$$

Thickness is a logarithmic function of time.

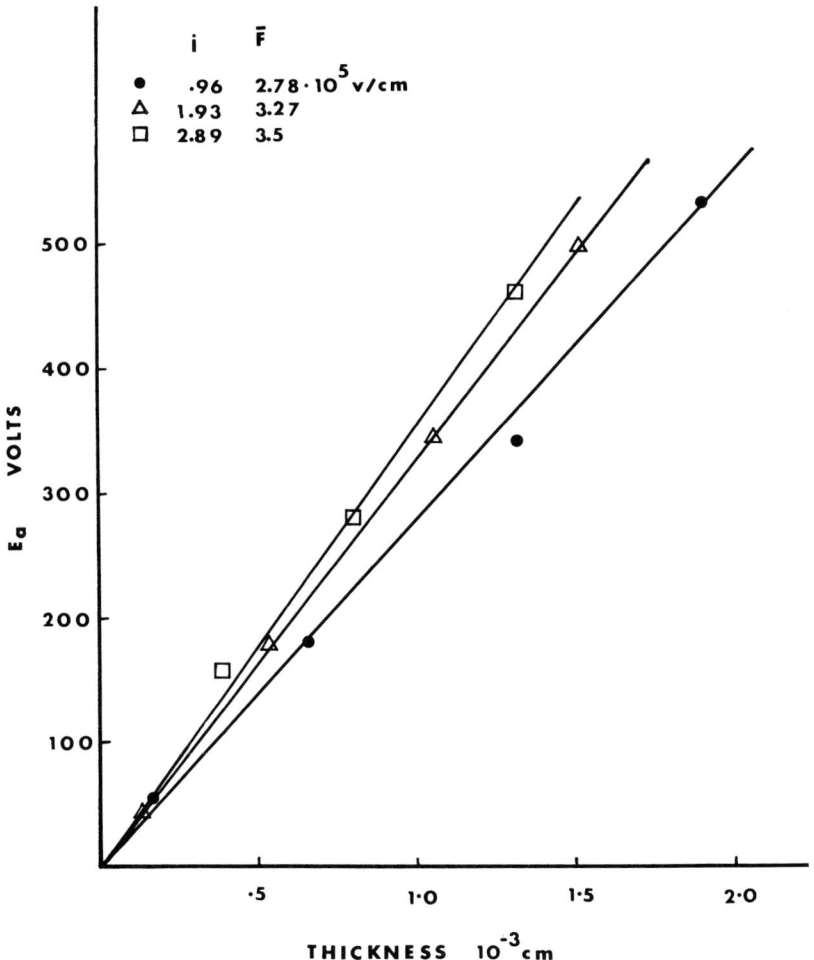

Figure 4. Anode potential vs. thickness of baked film at constant j in oil-modified polyester.

What values of an electrical field exist during the electrodeposition of polymers, and how are they determined? Field is defined as E_a/δ (*11*); by plotting E_a vs. δ (Figure 4), one obtains a differential field. In Figure 4 these lines intersect at the same point, indicating that voltage drop caused by the interfaces is independent of current and is small in comparison with a voltage drop across the film thickness. In these plots, the thickness of the baked film is used instead of the wet films, assuming that the error introduced is such that a constant can be introduced. The fields obtained are such that when a given *j* was plugged into the computer, it calculated the hyperbolic sines relation according to Equation 6. The plots are given in Figure 5 for a low molecular weight system and in Figure 6 for a high molecular weight system. For low molecular

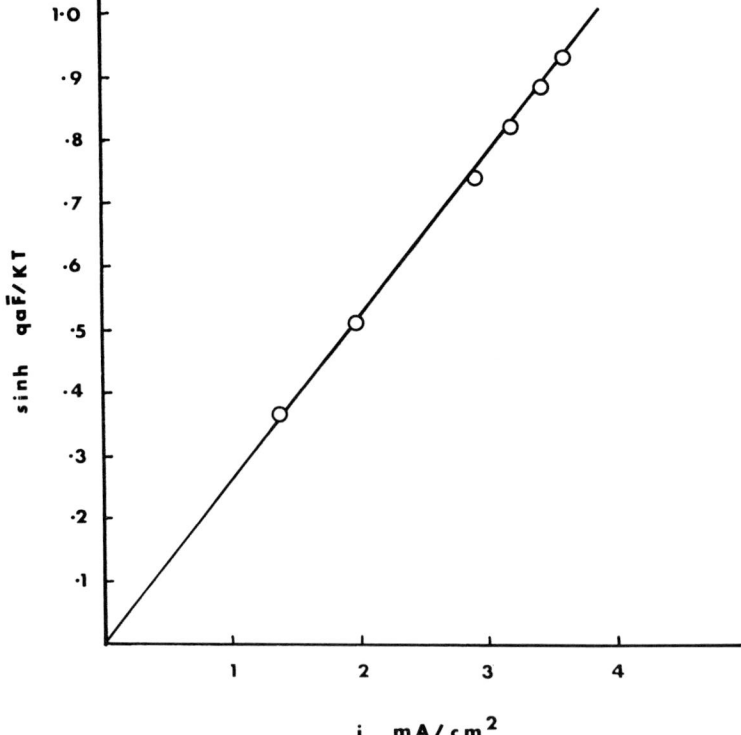

Figure 5. Hyperbolic sine function vs. ionic current density in an epoxy ester

weight systems, sinh varied between 0.37 and 0.9 since the $q a \vec{F}/kT$ varied between 0.3 and 0.8 or $\vec{F}$ between 1.2 to 2.8×10^5 volts/cm. However for high molecular weight systems, sinh changed from 5 to 25

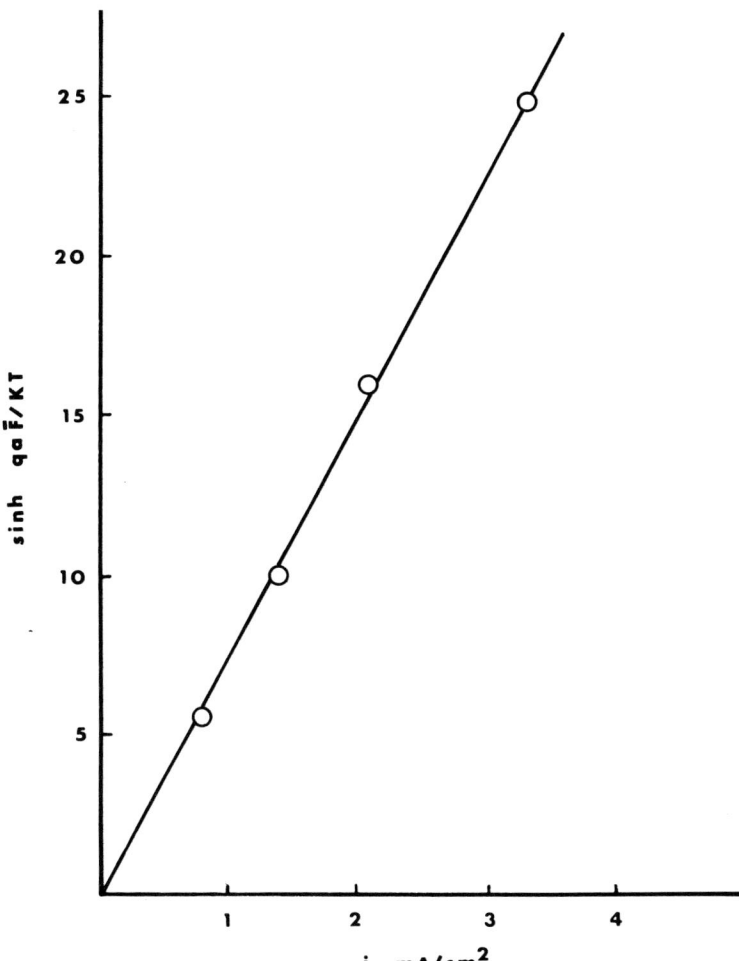

Figure 6. Hyperbolic sine function vs. ionic current density in an oil-modified polyester

since $q a \vec{F}/kT$ changes from 2–5 or $\vec{F}$ from 3 to 5 × 10^5 volts/cm. From the computer data one can calculate qa. If q is assumed to be +1, then in the low molecular weight system, a is 8 A and in the high system is 35 A. The distances between the carboxylic groups of the polymer in these two systems (*12*) are 10 and 20 A for low and high M, respectively. This would indicate that H^+ ions will have enough energy to move from one carboxylic group to the other.

From the above equations, one expects that thickness would vary linearly with time at $j = $ constant; this is shown in Figure 7. At $E_{app} = $ constant, for low to moderate fields, $\delta = \sqrt{t}$ at 4 seconds and longer,

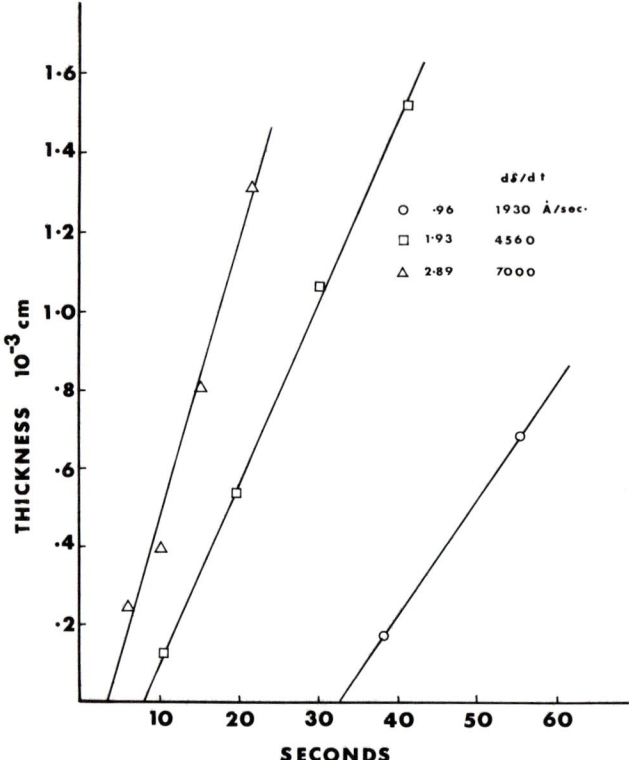

Figure 7. Thickness of baked films vs. time in an oil-modified polyester at constant current density

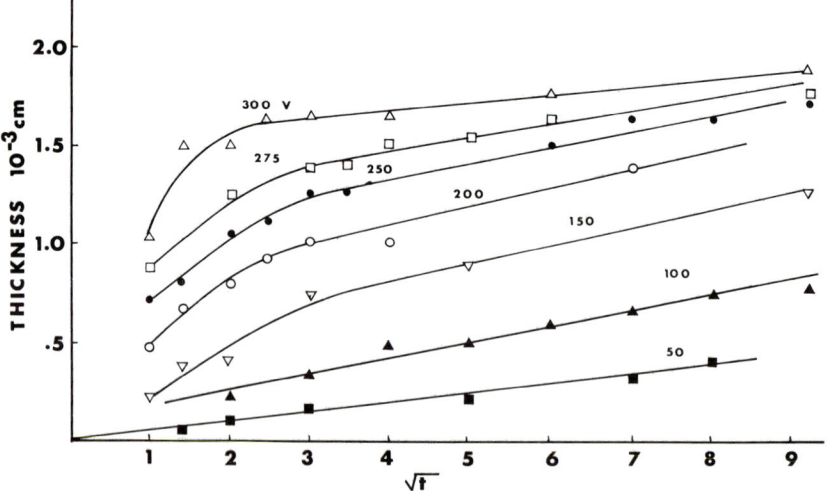

Figure 8. Thickness of baked films vs. $\sqrt{time}$ in an epoxy ester at constant applied voltage

which is shown in Figure 8. At the beginning of electrodeposition and at high E_{app} where $q a\vec{F}/kT > 1$ there is deviation from the $\sqrt{t}$ law.

Table I shows the field after successive times at $E_{app} = 350$ volts in a high molecular weight system. This shows what type of growth can be expected. For short times qaF is $\sim 6\ kT$, which means that thickness would vary linearly with log t, and at longer times where $q a\vec{F} = 1$–2 kT, the $\sqrt{t}$ dependence would be approached after 9 to 16 seconds. This is revealed in Figures 9 and 10.

Table I. 350-volt Oil-Modified Polyester Coating

Time, sec	j, ma/cm^2	Field, volts/cm $\times\ 10^5$	$q a\vec{F}/kT$
0.27	44.8	6.37	6.13
0.826	15.2	5.25	5.05
1.14	9.97	4.81	4.63
2.25	4.40	3.96	3.81
4.0	1.74	3.00	2.89
16.0	0.76	2.16	2.08
36.0	0.67	2.02	1.94
81.0	0.49	1.72	1.66

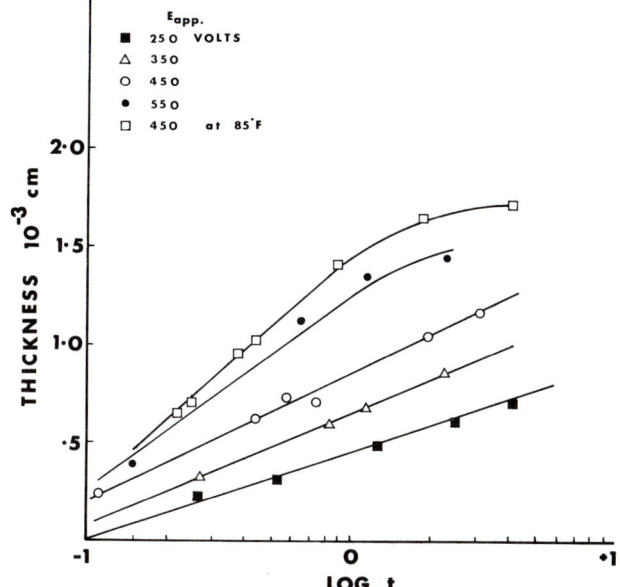

Figure 9. Thickness of baked film vs. log time in an oil-modified polyester at constant applied voltage. Short electrodeposition times.

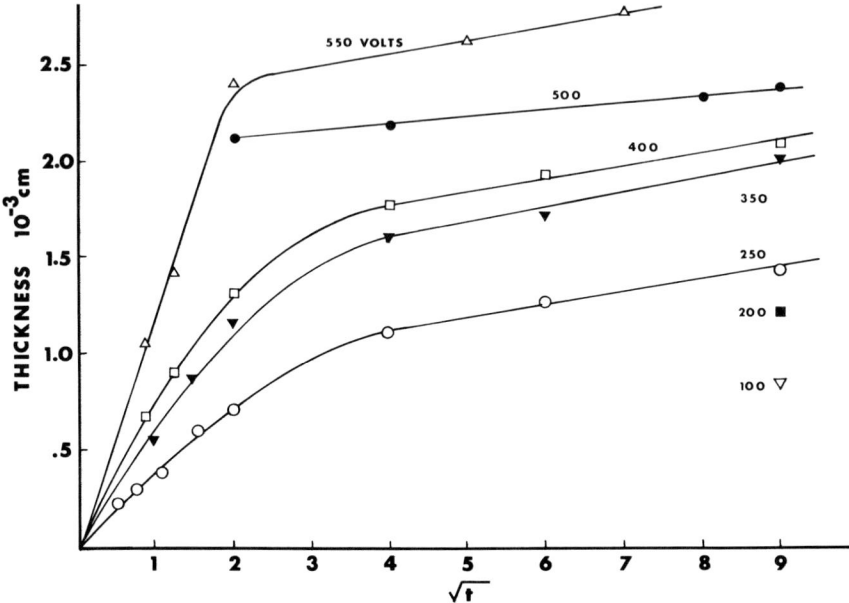

Figure 10. Thickness of baked film vs $\sqrt{time}$ in an oil-modified polyester at constant applied voltage

At constant time, thickness increases linearly with E_{app}, as shown for an oil-modified polyester system in Figure 11. The slopes and intercepts of these lines differ for short (4 seconds) and long (81 seconds) times because of the differences in the two growth processes. The calculation is based on the differences between electrical and thermal energies.

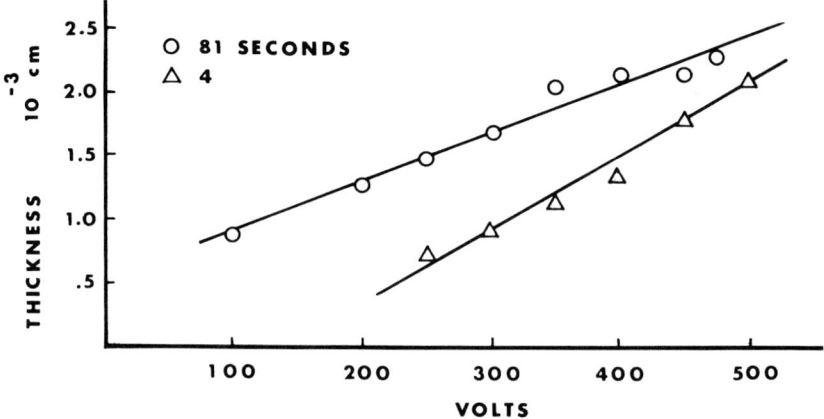

Figure 11. Thickness of baked film vs. applied voltage in an oil-modified polyester at constant electrodeposition time

How is it known that kT is not much larger than assumed? To clarify this, the temperature of the substrate was measured during electrodeposition with a Thermistor (DynaSense electronic thermometer model 8390-3) which was in thermal but not electrical contact with the substrate because one does not want to measure EMF caused by passage of current through it but the EMF caused by thermal changes. The data for the highest possible ΔT change are shown in Figure 12. This figure shows that in the stirred system ΔT changes only by a few degrees C, which would make a negligible contribution to the calculations.

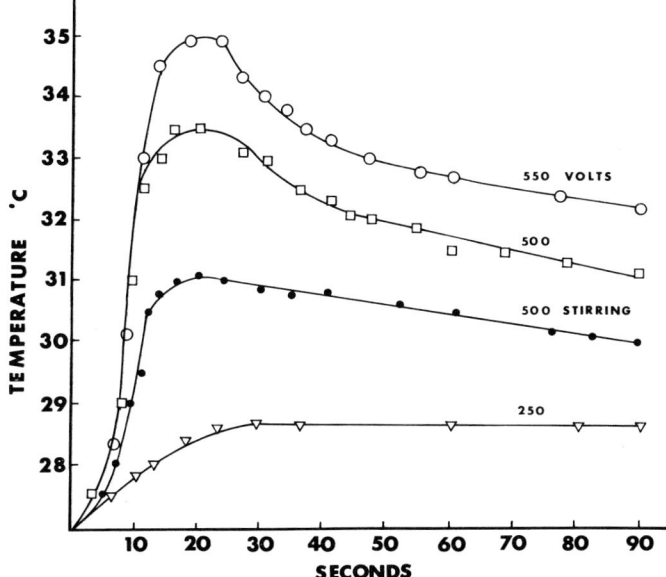

Figure 12. Temperature of substrate vs. electrodeposition time in an oil-modified polyester

Coulombic Efficiency. Coulombic efficiency, CE, is defined as mg of baked film per coulomb passed. At $j =$ constant, where $\vec{F}$ is constant, one has CE constant, or mg vs. coulombs are linear, as shown in Figure 13. The slope of these lines is the CE, from which one can calculate M/n—i.e., the electrochemical equivalent weight.

At $E_{app} =$ constant, the coulombic efficiency is also independent of the electrodeposition time but increases with an increased applied voltage as shown in Figure 14. If the polymer system had a narrow molecular weight distribution and would deposit as a stoichiometric compound, such variations in the equivalent weight would not be possible. However, we are dealing with systems which have neither a narrow molecular weight distribution nor are they deposited as "pure" compounds. (Chemi-

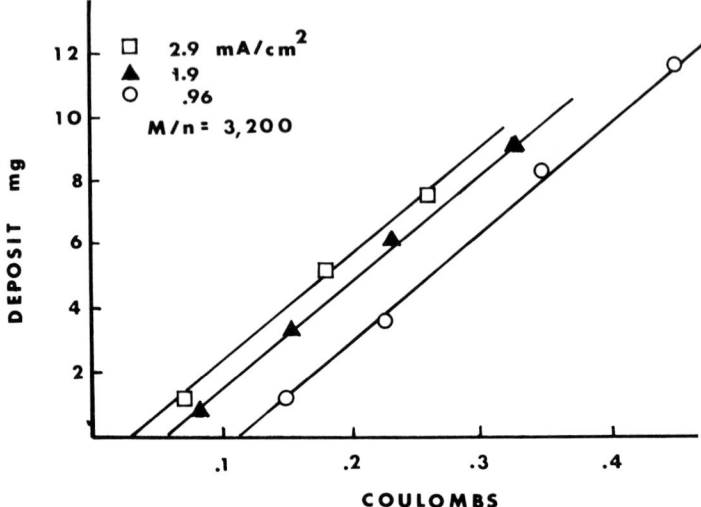

Figure 13. Weight of baked film vs. coulombs passed in an oil-modified polyester at constant current density

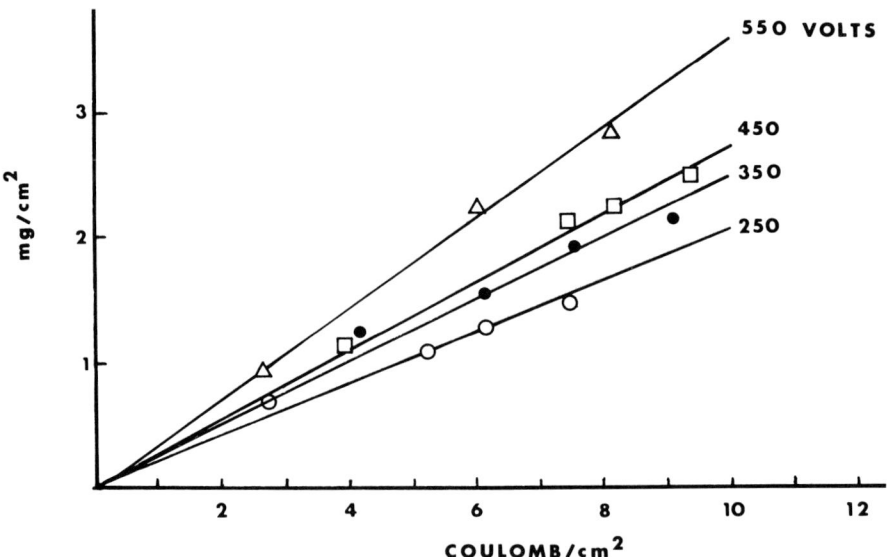

Figure 14. Weight of baked film vs. coulombs passed in an oil-modified polyester at constant applied voltage

cal analysis of wet films showed that they also contain trapped counterions.)

Resistance. Anode potential (E_a) represents the sum of voltage drops across the metal–film interface, across the film thickness, and across the film–paint interface. E_a plus the IR drop through the paint emulsion and the cathode–paint interface equals E_{app}. Since most of the voltage drop occurs across the film itself, the ratio of E_a and current density passing to deposit a film at a given time represents resistance, R, of wet, growing film and is given in Ω/cm^2.

As pointed out earlier at $j =$ constant current R film would be a linear function of time since $\rho =$ constant. This is shown in Figure 15.

From known R and a thickness of baked film, the specific resistivities were calculated and were found to be 8.9×10^7 Ω cm and 8.8×10^8 Ω cm for the epoxy ester and the oil modified polyesters, respectively.

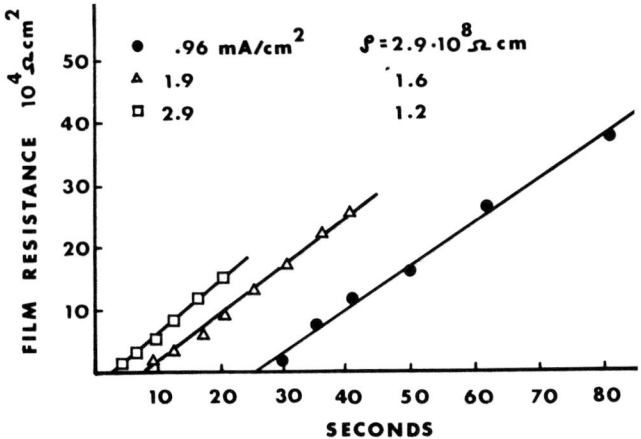

Figure 15. Film resistance vs. time in an oil-modified polyester at constant current density

However, at $E_{app} =$ constant, non-ohmic behavior should be expected at high fields. Non-ohmic conduction in polymer films was previously mentioned by Beck (13) and Cooke (14). Beck assumed that this behavior is caused by the presence of a space charge region, and Cooke assumed it was caused by the second Wien effect—i.e., field-induced dissociation of weak electrolytes. The fields during the electrodeposition of polymers are certainly high enough to cause the dissociation of deposited free acids or metallic soaps. Since in the present work the Wien effect and its contribution to the non-ohmic conduction in the films were not measured directly, R was calculated only for the epoxy ester system, where low to moderate fields were found—i.e., where Ohm's law is valid. Specific resistivities of these films are given in Table II.

Table II. Specific Resistivity[a] for Epoxy Ester

Voltage, volts	Time, sec.	Ω cm $\times 10^7$
200 volts	4	3.2
	9	7.6
	49	9.6
275 volts	4	6.0
	9	9.0
	49	10.8

[a] Resistivity increases with time and E_{app} in this system.

Summary

During the electrodeposition of polymers at constant voltage, the current density is a function of the hyperbolic sine of the electrical field. Film growth at the beginning of electrodeposition follows the logarithmic time law. In this region the film resistance is non-ohmic in behavior. This is the region of high electrical field strength.

As the film thickness increases, the field strength across it decreases, and the growth of the film follows the $\sqrt{t}$ law. In this region the film resistance follows Ohm's law. This is also a region of moderate to low field strength.

At constant current, the thickness and film resistance vary linearly with time. Coulombic efficiency and specific resistivity are constant since the electrical field is maintained constant during the deposition. Coulombic efficiency increases, and film resistance decreases with increased current density.

Acknowledgment

The author thanks C. Higginbotham for his skillful experimental assistance. Grateful acknowledgment is given to P. Pierce, N. Frick, and M. Wismer for their many helpful and challenging discussions.

Literature Cited

1. Fink, C. G., Feinleib, M., *Trans. Electrochem. Soc.* (1948) **94**, 309.
2. Tawn, A. H. R., Beery, I. R., *J. Oil Colour Chem. Assoc.* (1965) **48**, 790.
3. Beck, F., *Farbe Lack* (1966) **72**, 218.
4. Finn, S. R., Mell, C. G., *J. Oil Colour Chem. Assoc.* (1964) **47**, 219.
5. Netillard, J. P., *Double Liaison* (1970) **17**, 225, 233.
6. Saatweber, D., Vollmert, B., *Angew. Makromol. Chem.* (1969) **8**, 80; (1970) **10**, 143.
7. Finn, S. R., Hasnip, J. A., *J. Oil Colour Chem. Assoc.* (1965) **48**, 1121.
8. Beery, I. R., *Paint Technol.* (1963) **27**, 12.

9. Bockris, J. O'M., Reddy, A. K., "Modern Electrochemistry," Vol. I, p. 391, Plenum, New York, 1970.
10. Cabrera, H., Mott, N. F., *Rep. Progr. Phys.* (1948-49) **12**, 163.
11. Young, L., "Anodic Films," Academic, New York, 1961.
12. Pierce, P., private communication (1970).
13. Beck, F., *Ber. Bunseng., Physik. Chem.* (1968) **72**, 445.
14. Cooke, B. A., *Paint Technol.* (1970) **34**, 12.

RECEIVED May 28, 1971

11

Variables Affecting the Kinetics of Polymer Electrodeposition

W. B. BROWN and G. A. CAMPBELL

General Motors Research Laboratories, Warren, Mich. 48090

> *Investigation of the kinetics of polymer electrodeposition has shown that Joule heating of the deposited film is a major factor to be considered. As the film resistance increases, film heating increases to a maximum, then decreases. The point of maximum heating is the critical stage for film rupture. Rupture was caused by boiling of the water trapped in the film. If heating is sufficient to boil the liquid, rupture is self-sustaining. If not, rupture is suppressed, even at higher voltages, once the critical region is passed. Scanning electron microscope photographs of the film after rupture show smooth cratered areas surrounded by a distinct ridge. These are not present in coatings deposited below the rupture voltage. They appear to be the base of bubbles created when localized areas boiled.*

Early research into the kinetics of electrodeposition of paint was rather incomplete. Simple kinetic schemes were proposed, where only the latter stages of deposition were considered (1). Observed deviations from the kinetics were described as being the result of "compaction" of the film. Other investigators confined themselves to studying phenomena associated with the deposition process, but they failed to deal with the overall process. Conflicting reports in the early work can be traced to the use of different types of resins. Within the past few years, papers have appeared which take a broader and more realistic view of the deposition process. Beck (2, 3, 4) has investigated several factors affecting the deposition, including boundary layer reactions and charge migration within the deposited film. Saatweber and Vollmert (5, 6, 7) have treated the problem from a materials standpoint and have shown effects on stability of the paint suspensions, coulombic yield, and film resistance governed by the type of resin used. Yeates (8) in his book, "Electropainting," has summarized the effects of the many variables on the deposition as well as the more practical aspects of appearance and performance of the coated parts.

The purpose of this investigation was to perform an overall study of the kinetics and the factors influencing the kinetics of the electrodeposition of paints. The intent was to supply data for a dynamic computer simulation of the deposition process. The observations made in this investigation have added to the understanding of the process, and the more important factors are discussed.

Experimental

The resins used were commercial. One was a maleinized linseed oil, pigmented and unpigmented, which was dispersed in distilled water with diethylamine to bring the pH to a value of 7. The other, a clear resin, was a maleinized type, similar in nature to that used by Rheineck (9). Depositions were conducted in either a 4-liter glass beaker or in a 55-gal stainless steel tank equipped with pump, filters and a separate cathode of an area at least four times that of the panel.

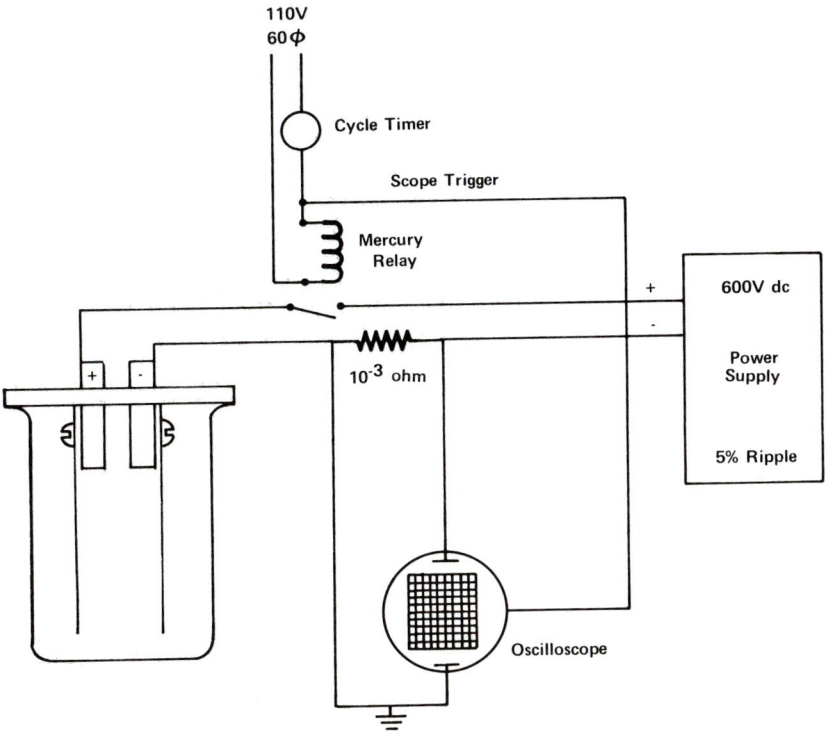

Figure 1. Circuit diagram for current monitoring

The panels used were Bonderite EP-1 or Bonderite 40 coated (available from the Parker Rust Proof Div., Hooker Chemical Co.). They were masked with vinyl tape to give a controlled surface area of 12 sq in directly facing the cathode. The panels were mounted in a holding fixture to ensure adequate electrical contact and to maintain the relative positions of anode and cathode at 1.75 in during the deposition.

The power supply was a Darrah 600-volt dc power supply rated at 50 amps. Ripple was measured at 5% throughout the range of voltages. During the deposition, current flow and voltage were monitored with the system shown in Figure 1. Either a Tektronix 564B storage oscilloscope or a Photovolt Varicord model 44 strip chart recorder was used to monitor current. The salient features of this system are a mercury relay switch which allows instantaneous switching of current at full operating voltage. The relay was also used to trigger the oscilloscope scan to allow current measurements during the initial deposition times. A cycle timer serves to switch the mercury relay and allowed deposition times to be reproduced within 2 seconds. Current measurements were made across a low resistance (10^{-3} ohm) shunt. With this system, voltage rise time to the operating voltage was on the order of 1 msec.

Film temperatures were measured by using a sheathed copper–constantan thermocouple whose overall diameter was 15 mils. The couple was mounted to the grounded side of the circuit to prevent overloading the amplifier at operating voltage. Electrical contact was made between the panel and the thermocouple sheath to allow the film to build up on the thermocouple as well as on the panel. In this way, more accurate film temperatures would be obtained.

The scanning electron microscope was used to investigate surface

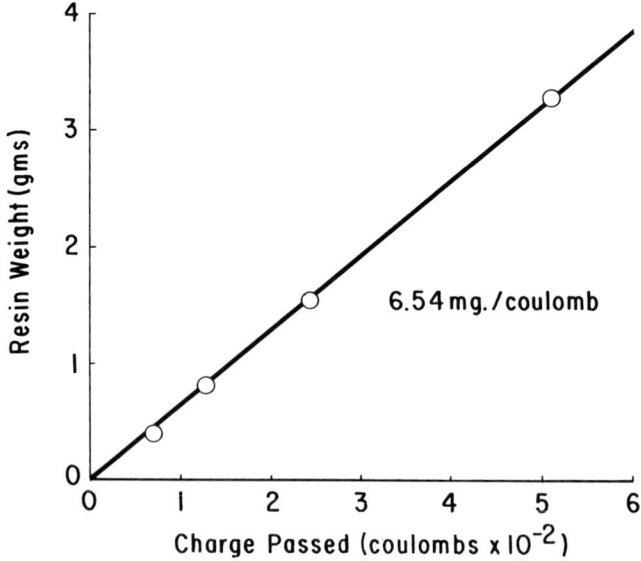

Figure 2. Coulombic yield curve

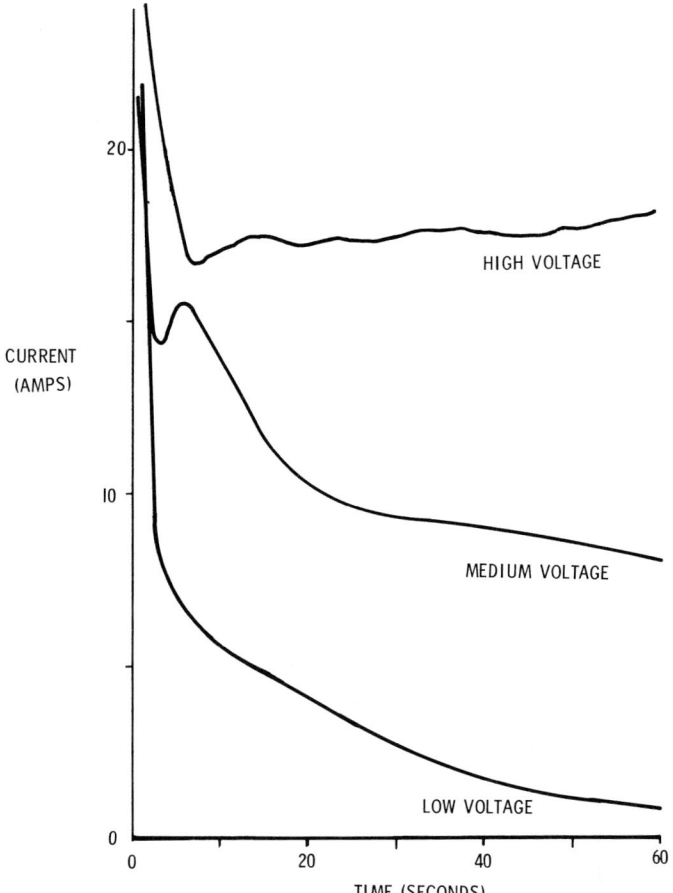

Figure 3.   Current variations with voltage

morphology. To retain the surface structures after deposition, the panels were not baked but allowed to air dry for about one week. Small rectangles were cut from the panels, and drying was completed at room temperature in a vacuum oven.

Results

The deposition kinetics were followed by observing current flow as a function of time. Figure 2 illustrates that for the resin under study a linear relationship exists between the amount of resin deposited and the current passed. Pigmented materials show a deviation from linearity and some voltage dependence on the amount of material deposited. Only a slight effect was observed with the pigmented material, however, and for our purposes this difference will be ignored.

Typical plots of current with time are shown in Figure 3. The lower curve illustrates the characteristic current plot at low voltages. The curve is smooth, showing a rapid increase in resistance within the first 2 to 3 seconds, then the current curve begins to level off. The initial current rise is so rapid that the speed of the recorder is insufficient to measure the maximum current draw. Oscilloscope measurements are necessary in this region of the curve to determine the current accurately.

The middle curve was recorded at an intermediate voltage and shows a similar increase in resistance in the first 2 to 3 seconds. The current plot then proceeded through a minimum, followed by a maximum. A similar plot appeared in a report by LeBras (*10*). The current curve then tended to level off under more or less steady conditions. The upper curve taken at a high voltage shows the initially increasing resistance. The curve went through a minimum, then increased indefinitely in an apparently uncontrolled manner. As shown by Rheineck (*9*), this uncontrolled rise in the current is characteristic of film rupture.

Figure 4 shows the panels which were deposited at these three voltages. The first panel shows a smooth, flaw-free surface, and would be

Figure 4. Panels electrodeposited at low (left), medium (center), and high (right) voltages

satisfactory for use. The second panel shows a moderate amount of coarseness which would require sanding to provide a satisfactory coating. Finally, the last panel shows extreme roughness and on close examination would show craters which extend through the coating to the metal surface. In this extreme case, corrosion resistance as well as appearance is affected.

During observations of film rupture, excessive bubbling was noted, and an increase in bath temperature of 5 to 10 degrees was observed for each panel deposited. The similarity in appearance to boiling led to the mounting of a thermocouple on the panel and the simultaneous recording of both current and temperature. Figure 5 shows that the onset of film rupture as indicated by current rise is accompanied by a temperature increase within the film to about 120°C, the boiling point of the bath liquid.

Figure 5. Current and temperature during film rupture

The reason for the uncontrolled increase in current is the physical removal of already deposited film by the boiling action within the coating. A greater area of bare panel is exposed, and thus the resistance is decreased. The amount of current flowing is determined by the amount of exposed panel area and is not predictable from a kinetics standpoint. As suggested by Rheineck (9), the linear relationship between deposited resin and current begins to fall off rapidly when rupture occurs. In fact, it was proposed (9) that the deviation from linearity of the coulombic yield curve be used to determine the rupture voltage.

The catastrophic nature of the film rupture was significant in determining the mechanism of this failure. Observations of film rupture led to findings which were difficult to explain. One observation was that film rupture did not occur until most of the film build was complete. Owing to the shape of the deposition curve, most of the material is deposited in the first few seconds. From electrical current measurements, about 60% of the film was deposited before rupture occurred. When the voltage is raised 200 to 300 volts above the rupture voltage, the shape of the initial part of the deposition curve is changed only slightly in shape. Rupture always seemed to be associated with the appearance of the inflection point in the current plot as shown in Figure 3.

If one considers the respective voltage drops between the anode and cathode, and treats them as both current controlling resistances and as resistive heating elements, the relationship to film heating can be developed. Initially, the current is limited by the bath resistance since this is the largest resistance between the anode and cathode. In this region, the heat is generated over a large volume of the liquid bath. As the film begins to increase in thickness, its resistance increases, further limiting the current flow. At constant voltage according to Ohm's law, the current can be calculated from the following relationship:

$$E = IR$$

where E = voltage (constant)
I = current
R = apparent resistance of the entire system

Simultaneously, the heat generated in the system:

$$W = I^2R$$

where W = watts of heat

When the resistance associated with the film equals that of the bath, maximum heat generation within the film is taking place.

The plot of Figure 6 demonstrates this. At 200 volts, the bath resistance is typically 25 ohms, and the calculated current flow is 8 amps. As the resistance of the film increases, the total current flow is reduced, but the heat being built up in the film is increasing. At 25 ohms film resistance, half of the heat is restricted to a film of roughly 2.0 mils. At this point, the film rupture becomes critical. If rupture occurs, the film resistance decreases, and the system would move to the left on the heat curve of Figure 6 allowing rupture to continue. If film rupture is prevented from occurring at this point, the system progresses to the right on the heating curve, and rupture does not occur since a more stable condition has been reached.

From Figure 6, it can be generalized that the maximum heat generated within any film occurs when the apparent film resistance equals

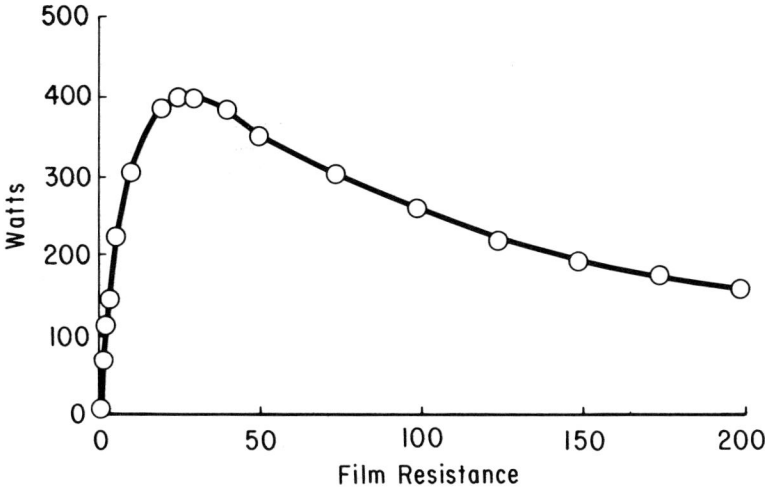

Figure 6. Film heating during electrodeposition

that of the bath. Even when film rupture does not occur, rather high temperatures are reached within the film. To determine the effects of temperature on the deposition, the bath conductivity was measured at various temperatures. The results are shown in Figure 7. Apparent film resistance may be considered to have a similar temperature dependence since conduction through the film is most likely ionic in nature.

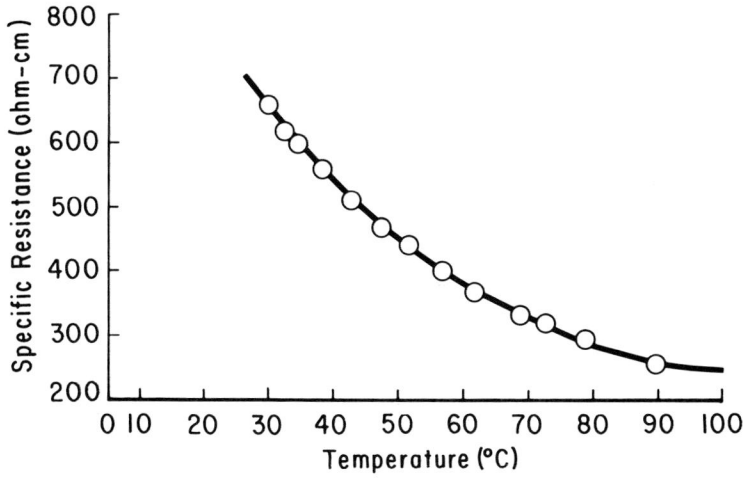

Figure 7. Bath resistance for various temperatures

The effect of rising film temperature is shown in Figure 8. As the temperature rapidly approaches its maximum, the increased conductivity produces a maximum in the current curve. As was previously stated, the maximum heating takes place when the apparent film resistance and the bath resistance are equal. At this point, the total resistance of the system is twice the initial resistance. The maximum in the current curve would thus occur at half the initial current. Because of the sluggishness of the strip chart recorder, oscilloscope measurements of initial current were taken, and in all cases the current at which the inflection occurred was half the initial current. After passing the critical region, the film temperature decreases until at the end of a 1-minute deposition, the film temperature almost equals that of the bath (22°C).

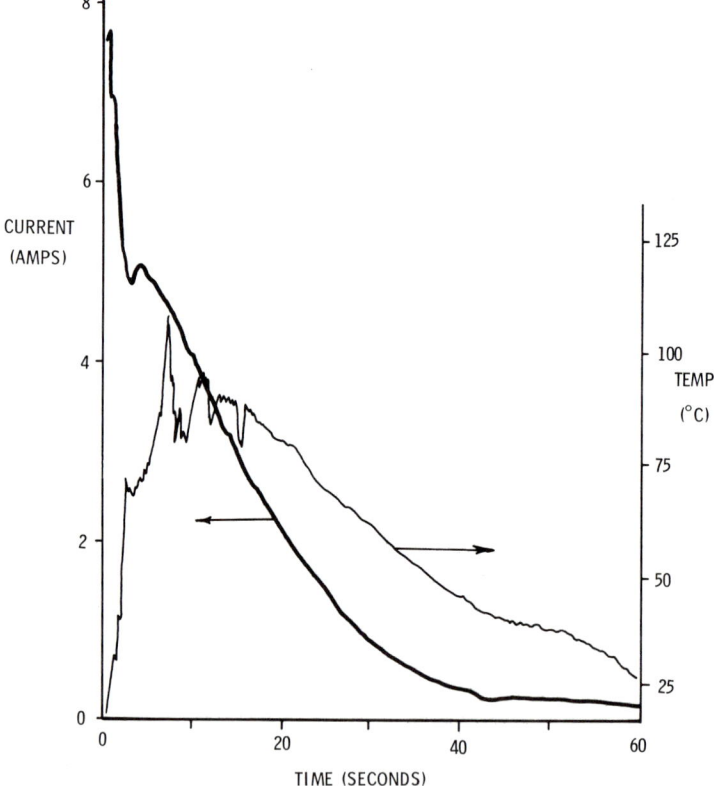

Figure 8. Current and temperature at medium voltage

This temperature increase during the deposition also has the effect of developing a smoother film. During low voltage depositions, an extremely coarse coating resulted. Baking improved the appearance, but the optimum appearance resulted only with higher voltage depositions. The heat buildup within the film appears to improve the flow of the film to a certain extent and to allow the deposited particles to knit.

In a detailed study of surface morphology of these coatings, some of the thermal effects became apparent. Figure 9 shows some changes in the surface under different deposition voltages, as observed with the scanning electron microscope. At 200 volts, a porous coating was developed. The bridges between coagulated clumps were numerous, but little flow had occurred. At 300 volts, a somewhat smoother surface was evi-

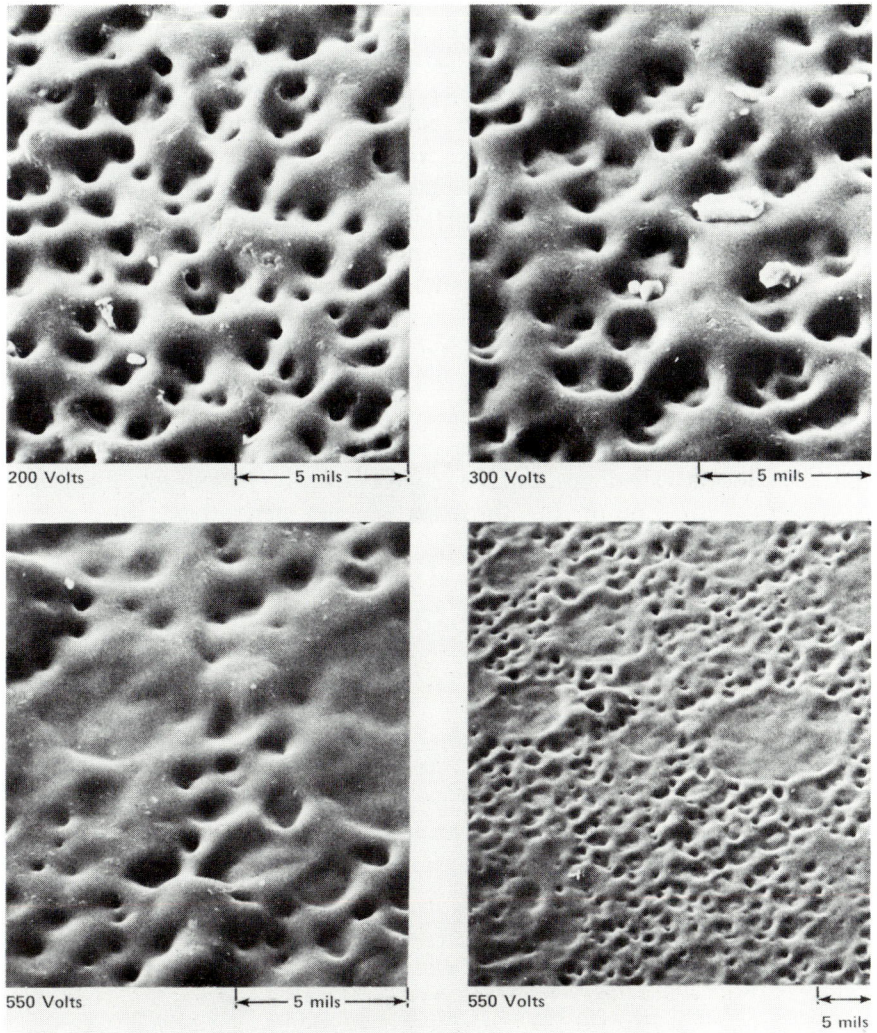

Figure 9. Surface structures of electrodeposited films

dent. Spaces between the clumps had begun to level, and the bridges between particles had increased in size. At 550 volts, slightly above the rupture voltage, the surface was smoothed out considerably. There was, apparently, a new structure present when the rupture voltage was exceeded. These were flat, well-knit areas surrounded by a well-defined rim. These structures are currently ascribed to the site of local film rupture. They appear to be the base of a resin bubble formed by the vaporizing bath liquid. Adjacent to some of these structures, one can find lumps of material which appear to have been the dome of the bubble.

Conclusions

From our studies of the kinetics of electrodeposition, it has become apparent that a major factor to be considered is heating within the film during the deposition. In the extreme case, the rise in temperature is sufficient to produce boiling of the liquid contained in the film and film rupture. Even at lower voltages, the film temperature increase has an effect on the appearance of the deposited coating.

Treating the deposition bath and the deposited coating as series resistances, the Joule heating of the film reaches a maximum when the bath and apparent film resistance are equal. At this point, film rupture becomes critical. If rupture occurs, it is self-sustaining. If rupture is prevented, the film temperature decreases, and rupture cannot take place. This may explain, in part, the observations of Brewer (11) that after the initial 70 seconds of deposition, the voltage can be raised above the rupture voltage, thus improving throwing power. After the critical region in heating is passed, the system can tolerate a higher voltage without rupturing, owing to the decreased current flow at the higher film resistance.

Acknowledgments

The authors thank A. C. Ottolini and T. P. Schreiber for obtaining data on the electron probe and for SEM microphotographs of the surface of the deposited resins. The help of R. L. Grimshaw was appreciated in panel deposition and in tank maintenance.

Literature Cited

1. Finn, S. R., Hasnip, J. A., *J. Oil Colour Chem. Assoc.* (1965) **146**, 1121.
2. Beck, F., *Ber.* (1968) **72**, 445.
3. Beck, F., Pohlman, H., Spoor, H., *Farbe Lack* (1967) **73**, 298.
4. Beck, F., *Chemie-Ing.-Techn.* (1968) **40**, 575.
5. Saatweber, D., Vollmert, B., *Angew. Makromol. Chem.* (1969) **8**, 1.

6. *Ibid.* (1969) **9,** 61.
7. *Ibid.* (1970) **10,** 143.
8. Yeates, R. L., "Electropainting," Robert Draper Ltd., London, 1966.
9. Rheineck, A. E., Usmani, A. M., *J. Paint Technol.* (1969) **41,** 597.
10. LeBras, Louis R., *J. Paint Technol.* (1966) **38,** 85.
11. Brewer, G. E. F., Hines, F., "Preprints," *A. C. S. Div. Org. Coatings Plastics Chem.* (1970) **30** (2), 249.

RECEIVED May 28, 1971.

12

Dynamic Simulation of the Electrodeposition of Polymers

G. A. CAMPBELL and W. B. BROWN

General Motors Research Laboratories, Polymers Department, Warren, Mich. 48090

> *A dynamic mathematical model was developed for the electrodeposition of polymers using classical boundary layer theory. The dynamic changes in temperature and polymer concentrations were described for the bulk solution, boundary layer, paint film, and paint panel. The resulting equations were solved simultaneously on the hybrid computer. The data generated on the computer established the importance of the boundary layer in the process dynamics.*

The electrodeposition of polymers from aqueous systems has been studied extensively and is detailed by Yeates (*1*). That process, as described in this paper, deals with the changes which occur near the anode during anodic deposition of carboxylic-containing polymers. A computer simulation has been developed which reflects changes in resistances, current, and thermal transients for the deposition of paint films in a dc potential. Several alternatives have been suggested as to the electrical equivalence of the process (*2–7*). Many (*2–4*) propose that the system can be described by a linear resistance combination. However, several authors have detected nonlinear responses of resistance under changing potentials (*5–7*). The possible existence of a boundary layer has been described, but its importance as a resistance was discounted by the authors, (*5, 6*). The Nernst diffusion layer was described as being from 0.02 to 0.05 cm thick with the Prandtl boundary layer being an order of magnitude thicker. Beck *et al.* (*5, 6*) concluded that because the migration velocity was on the order of 10^{-2} cm/sec and the bulk velocity was from 10–100 cm/sec that the boundary layer has little effect. It is the opinion of the authors that the boundary layer is important in our system.

Model Development

There are five potentially important resistances in the deposition system as was suggested by Beck *et al.* (5) (Figure 1). The mathematical model developed in this simulation reflects the resistance character of the three major resistances: (1) R_B, the bath resistance; (2) R_D, the boundary layer resistance; and R_F, the paint film resistance. These three resistances in conjunction with the associated thermal transients were suggested by the deposition rate.

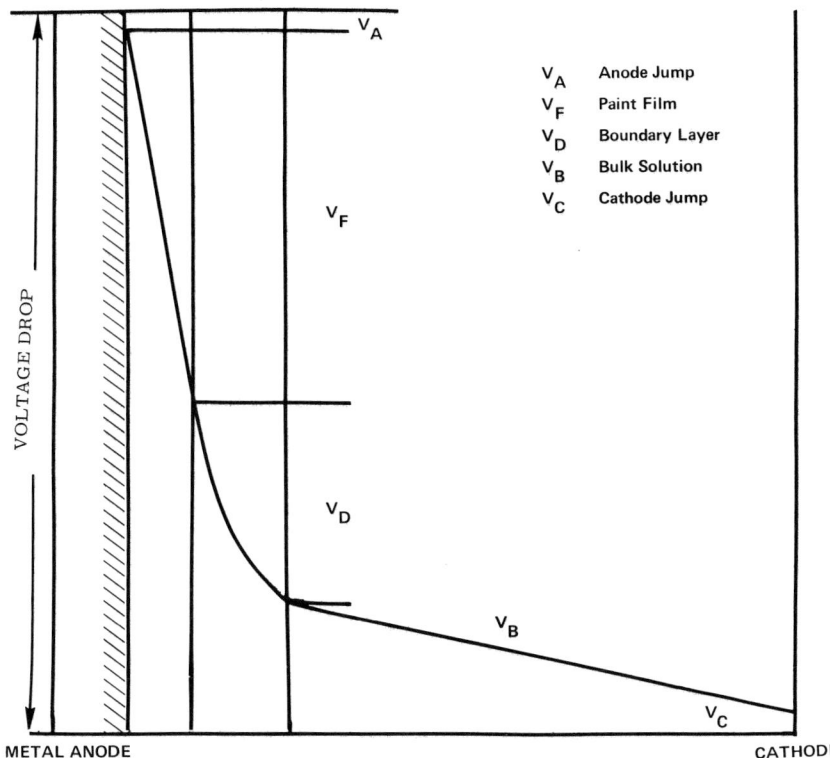

Figure 1. Potential drops for deposition system

Deposition Data

Typical current–time curves for clear solubilized resins are found in Figure 2. Figure 2 (top) is characteristic of a maleinized oil system and Figure 2 (bottom) is characteristic for high yield, good throw resins. The maximum in the curve (top) could be caused by a number of variables including temperature in the paint film or concentration and temperature changes in the boundary layer.

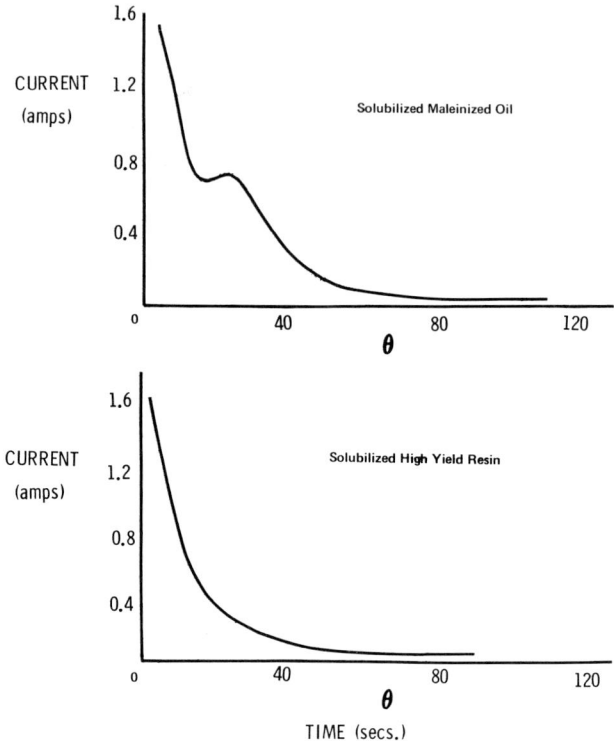

Figure 2. Current–time relationship

Cycling the voltage-produced current gives the curve in Figure 3. The results generally show a decrease in the transient current (the overshoot) with each cycle. Also, when pigment is added which increases the resistivity of the film, the transient decreased more rapidly with cycling.

Temperature is an important variable, as shown in Figure 4. Figure 4 (top) is representative of the current–time relationship for the blowing phenomenon of the maleinized oil system. Figure 4 (bottom) shows the temperature in the vicinity of the paint film. The temperature compounds with the increased current flow.

Finally, it was observed that agitation affects the current–time relation of the deposition process. The agitation was created using a mixer with deposition on a 3×4 inch panel area (Figure 5).

Bulk Properties

In this simulation the bath resistance was considered to be constant for any set of concentration and temperature conditions. The conductivity was best fitted by Equation 1:

$$K_c = AC + BC \sqrt{C} \tag{1}$$

The constants AC and BC were obtained using specific resistance data from Figure 6. The resistivity was also affected by temperature, as shown in Figure 7. The dual effect of concentration and temperature on conductivity was then incorporated into the model. The bath resistance can be calculated using the relation:

$$R_B = (1/K_c)(L/1\ cm^2) \tag{2}$$

All resistances computed in the model are based on a 1 cm² area.

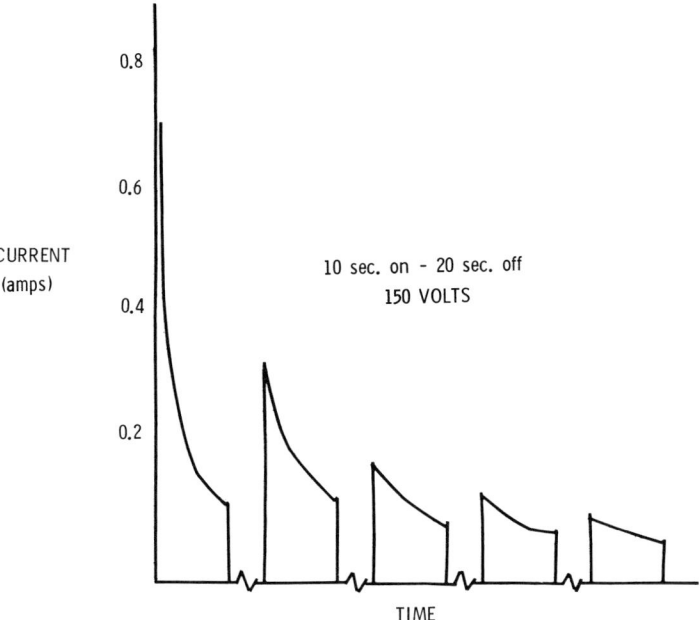

Figure 3. *Effect of cycled voltage*

Paint Film Properties

The thickness of the clear film was linearly proportional to charge flow. This is in agreement with reported data (8–10). It has been reported that the later portion of deposition on some substrates was not linear (8). However, as a first approximation for this simulation a linear relation was used. Therefore, the film thickness Y can be found by:

$$Y = A \int I(\theta) d\theta \tag{3}$$

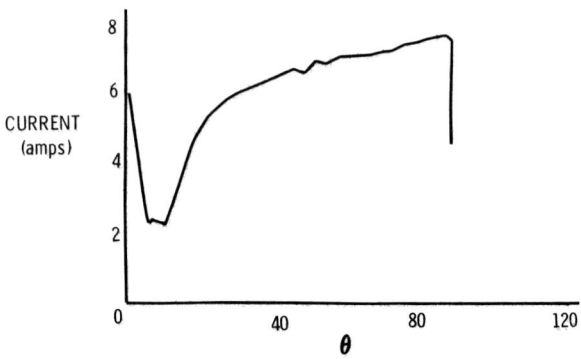
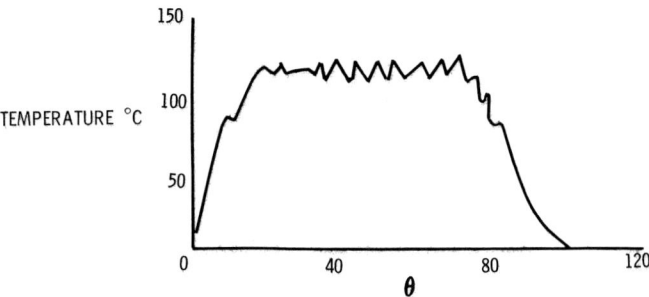

Figure 4. Relationship between current and temperature

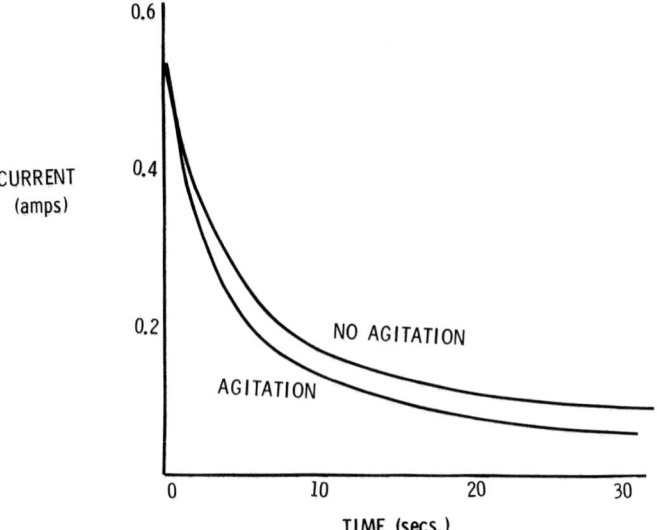

Figure 5. Effect of agitation

Although non-ohmic behavior (5, 7) has been reported, as a first-order model the resistance of the film will be considered to be only a function of film thickness:

$$R_F = BY \quad (4)$$

Most treatments of electrodeposition do not incorporate the thermal transients caused by I^2R dissipative losses. Because temperature changes were found to be associated with the blowing phenomenon (Figure 4), the thermal transients were also simulated. The temperature in the system can be shown to follow (11) the relationship:

$$(\partial T/\partial \theta)(\rho C_p) + \nabla \cdot Ni = Q \quad (5)$$

where
$$Ni = -k(\partial T/\partial Y) \quad (6)$$

For a one-dimensional heat transfer system, a combination of these equations leads to:

$$(\partial T/\partial \theta) = \alpha (\partial^2 T/\partial Y^2) + Q/\rho C_p \quad (7)$$

with
$$Q = (1/4.18)(I^2R)/Y \quad (8)$$

Boundary Layer

Although the boundary layer has been reported to be on the order of 0.01–0.1 cm (5), evidence indicates that it plays a role in the electrodeposition process dynamics. The increase in temperature near the film is suggestive of a boundary layer thermal resistance. If this boundary layer were very thin, the heat conduction rate would be large compared with the heat generation rate with little or no increase in temperature resulting. Also, the pulsed voltage experiments show trends which are consistent with boundary layer theory. The decreased overshoots of current with successive cycles suggests a dynamic resistance which decreases as film resistance increases. A concentration polarization in the boundary layer would account for this since it is proportional to the current flow. The pulsed voltage data do not wholly lend themselves to space charge theory (5, 6). The boundary layer, as used in this paper, is defined as the space where energy and mass are transported normal to the deposition surface by molecular transport. This would, therefore, include transport in the Prandtl boundary layer where the convective velocity vector is parallel to the anode, but the vector normal to the surface is zero because the flow is laminar regime.

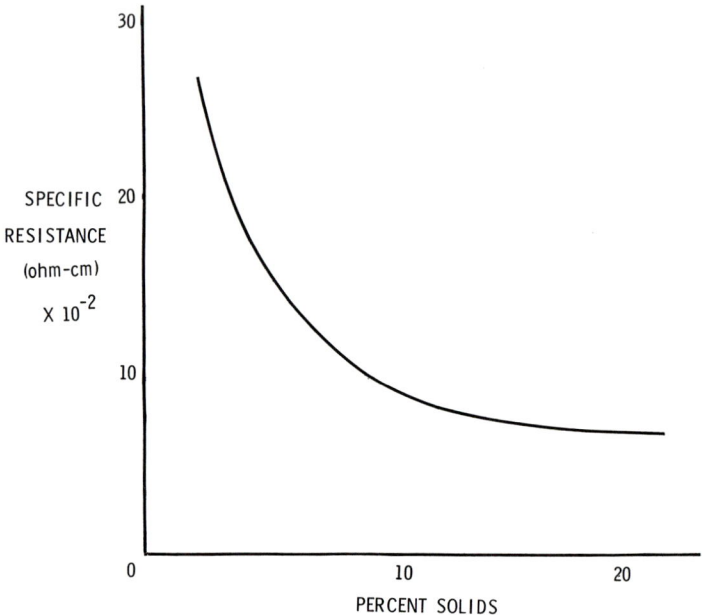

Figure 6. Effect of concentration on resistance

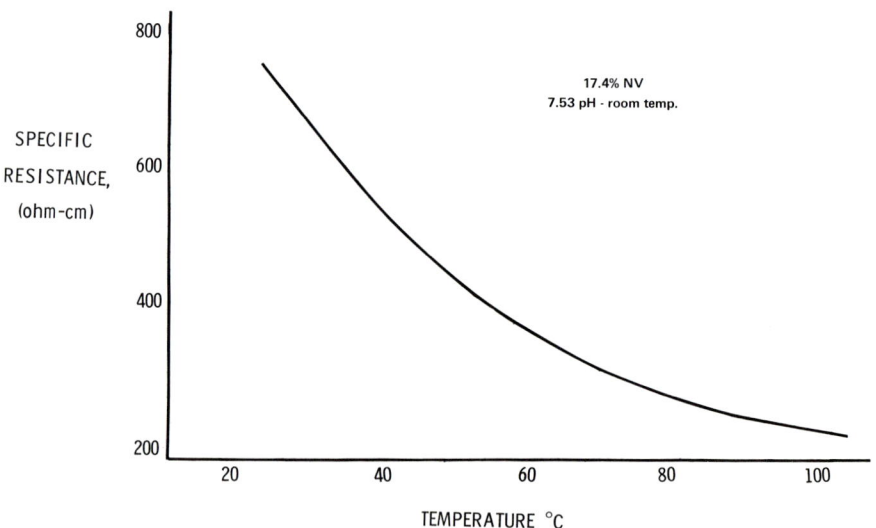

Figure 7. Effect of temperature on resistance

Assuming local electrical neutrality, the potential gradient in the diffusion layer is:

$$\nabla \Phi = -(I/K_c) \tag{9}$$

The time-dependent concentration can be represented by (11):

$$(\partial C_i/\partial \theta) + \nabla \cdot N_i = R_i \tag{10}$$

where the flux N_i is (12):

$$N_i = -D\nabla C_i - Z\mu_i F C_i \nabla \Phi + \vec{C V} \tag{11}$$

Using the Nernst-Einstein relation

$$D_i = \mu_i RT \tag{12}$$

we obtain

$$\partial C_i/\partial \theta + \vec{V} C_i = D_i \nabla^2 C_i + (ZFD/R)\nabla \cdot (C_i \nabla \Phi/T) + R_i \tag{13}$$

Since the generation, R_i, and the velocity vector in the direction of interest are zero, the assumption of a monodispersed system results in:

$$\partial C/\partial \theta = D\nabla^2 C + (ZFD/R)\nabla \cdot (C\nabla \Phi/T) \tag{14}$$

Expansion for the one-dimensional case gives:

$$\partial C/\partial \theta = D(\partial^2 C/\partial X^2) + (ZFD/R)$$
$$(\nabla C \nabla \Phi/T + C\nabla^2 \Phi/T - C\nabla \Phi \nabla T/T^2) \tag{15}$$

where:

$$\nabla^2 \Phi = (I/kc^2)(\partial C/\partial X) \tag{16}$$

The boundary layer resistance will then be

$$R_D = \Sigma \nabla \Phi/I \tag{17}$$

The temperature within the boundary layer would be described by:

$$\partial T/\partial \theta = \alpha(\partial^2 T/\partial X^2) + Q/\rho C_p \tag{18}$$

where:

$$Q = (\nabla \Phi I)/4.18 \tag{19}$$

The total resistance is then calculated:

$$R_T(\theta) = R_B + R_F + R_D \tag{20}$$

and the current may be calculated as:

$$I(\theta) = V/R_T(\theta) \tag{21}$$

Computer Solution

The preceding set of equations represents a first-order model of a very complex physical system. In a process such as this, with these complex nonlinear interactions, it is useful to use a hybrid analog digital computer (*13*) for the analysis. The computer used in this study was an Electronic Associates Inc. 690 hybrid computer.

The solution to partial differential equations with nonlinear terms and internal sources cannot be accomplished by classical separation of variables techniques. Therefore, a finite difference approach was used, with the grid shown in Figure 8. Standard difference equations were generated as described by Rogers and Connolly (*14*). The problem is then reduced to the solution of 14 simultaneous linear differential equations: five for heat transfer in the film, four for boundary layer heat transfer, four for concentration charges in the boundary layer, and one for the film thickness. These equations were solved on the analog computer, with much of the algebra being done on the digital computer. A hybrid task distribution is diagrammed in Figure 9.

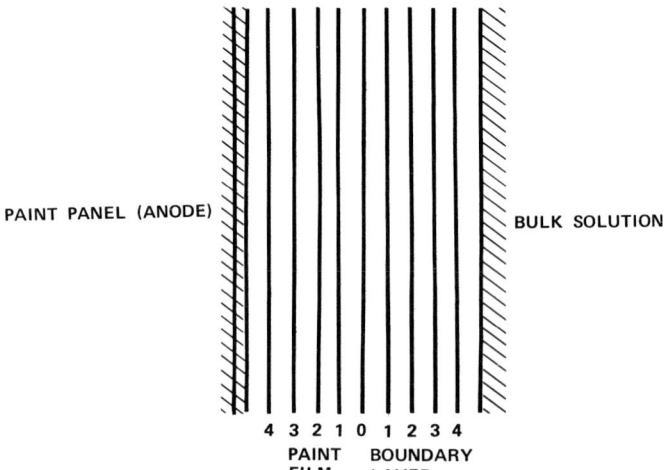

Figure 8. Finite difference section

Because of the nature of the deposition process, dynamic boundary conditions were used for the simulations. Since the material is being deposited at the anode, the concentration boundary conditions are:

$$X = X_{max} \quad C = C_0 \text{ (bulk concentration)}$$
$$X = 0 \quad C = 0 \quad V < 0$$
$$dC/d\theta = 0 \quad V = 0$$

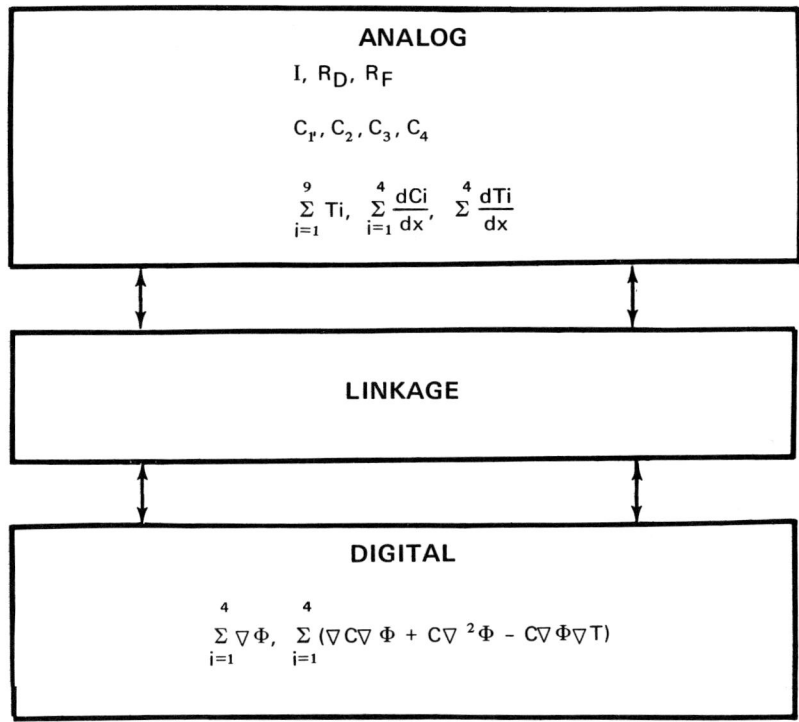

Figure 9. Hybrid task distribution

For temperature, the boundary conditions reduce to:

$$X = X_{max} \quad T = T_s \text{ (bulk temperature)}$$
$$Y = 0 \quad T = T_p \text{ (panel temperature)}$$

Information gained from the simulation is presented in Figure 10. These data were obtained using typical system parameters for the high yield resin. As would be expected since the potential gradient in the boundary layer is proportional to the current, the resistance increased as the voltage was raised from 200 to 300 volts. However, the resistance decreased when the voltage was increased to 400 volts. The apparent anomaly was explained by examining the concentration and thermal histories of the boundary layer. When the voltage was increased from 200 to 300 volts, the change in concentration had a greater influence than the thermal change—thus, the increased resistance. At 400 volts the thermal changes were the controlling factor, outweighing the concentration changes with the resulting decrease in resistance. These resistances constitute 10 to 20% of the total resistance in the early part of the deposition. The importance of the boundary layer in the thermal history

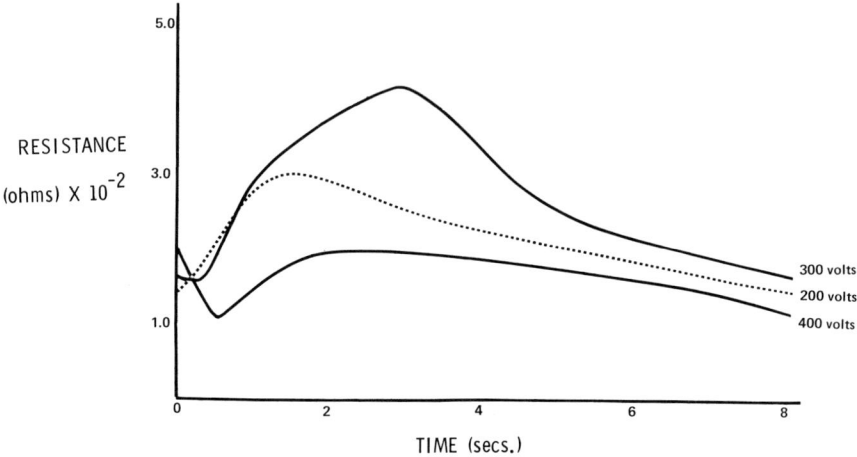

Figure 10. Boundary layer resistance as affected by voltage

of the system is seen in Figure 11. The computer was set up with typical deposition parameters at an intermediate voltage below the blow voltage. A temperature increase is generated, with most of the heat coming from the paint film; the heat then diffuses into the panel and boundary layer. When the boundary layer is removed, the heat can diffuse rapidly into

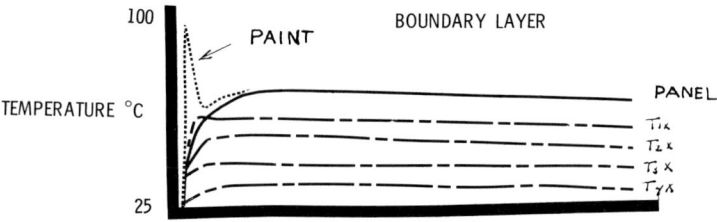

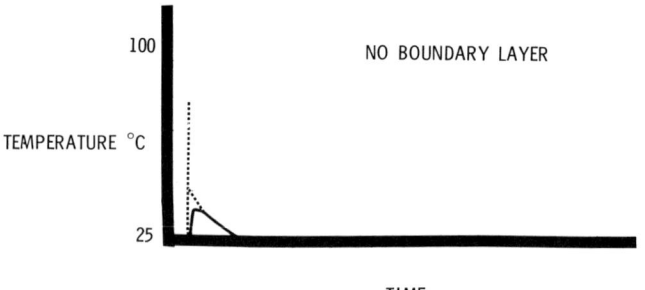

Figure 11. Temperature profile with and without boundary layer

the bulk; thus little temperature change is observed. The simulation with the boundary layer explains the experimental data presented in the paper on the heat generation in the film.

Simulation is now being expanded to compensate for changes in the resin film resistance caused by temperature and to incorporate potential dependent non-ohmic resistance in the film. The detailed second generation computer simulation will be presented in a subsequent publication.

Summary

Deposition data are presented which indicate that for this system the boundary layer plays an important role, first as a thermal insulator and secondly as a possibly important resistance in the early part of the deposition. A mathematical model is developed incorporating the three major process resistances with the associated heat transfer processes. The computer solution to this simulation is briefly described and computer data are presented which indicate that the changes in the boundary layer resistance are affected by both concentration depletion and temperature changes. This interaction causes the boundary layer resistance to go through a maximum as the voltage is increased from 200 to 400 volts. The model is being expanded in an effort to obtain a more complete solution.

Acknowledgment

The authors thank R. W. Valentine and R. Peugh for their assistance in developing the computer simulations.

Nomenclature

Symbol	Units	Variable
K_c	ohm^{-1} cm^{-1}	conductivity
AC	ohm^{-1} cm^{-1}	constant
BC	ohm^{-1} cm^{-1} (grams/cm^3)$^{-1/2}$	constant
C	grams/cm^3	concentration
R_B	ohms	bath resistance
L	cm	distance between anode and cathode
Y	cm	film thickness
A	cm/coulomb	coulombic efficiency
I	coulomb/sec	current
θ	seconds	time
R_F	ohms	film resistance
B	ohm/cm	constant
T	°K	temperature

Q	cal/cm³ sec	heat generation
ρ	grams/cm³	density
C_p	cal/gram °K	heat capacity
K	cal cm⁻¹ sec⁻¹ °K⁻¹	thermal conductivity
α	$K/\rho C_p$	thermal diffusivity
	volts	diffusion potential
Ri	gram-moles/cm² sec	rate of production
D, Di	cm²/sec	diffusion coefficient
Z		charge
μ	cm²-gram-mole/Joule sec	mobility
F	94,600	Faraday constant
V	cm/sec	velocity vector
R	Joules/gram-mole °K	gas law constant
R_D	ohms	boundary resistance
V	volts	voltage applied
R_T	ohms	total resistance

Literature Cited

1. Yeates, R. L., "Electro Painting," Draper, Teddington, 1966.
2. Finn, S. R., Hasnip, J. A., *J. Oil Colour Chemists Assoc.* (1965) **48**, 1121.
3. LeBras, L. R., *J. Paint Technol.* (1966) **2**, 38.
4. Holzinger, F., *J. Paint Technol.* (1966) **30** (9), 26.
5. Beck, F., Pohlman, H., Spoor, H., *Farbe Lack* (1967) **73** (4), 290.
6. Beck, F., *Ber. Bunsenges.* (1968) **72** (3), 445.
7. Cook, B. A., *J. Paint Technol.* (1970) **34** (1), 12.
8. Rheineck, A. F., Usmani, A., "Preprints," *ACS, Div. Org. Coatings Plastics* (1967) **27** (2), 193.
9. Rheineck, A. E., Usmani, A., *J. Paint Technol.* (1969) **41**, 597.
10. Geboy, J. P., Lahaye, J., *J. Paint Technol.* (1970) **42**, 371.
11. Bird, R. B., Stewart, W. E., Lightfoot, E. N., "Transport Phenomena," Wiley, New York, 1963.
12. Levich, V. G., "Physicochemical Hydrodynamics," Prentice-Hall, Englewood Cliffs, N. J., 1962.
13. Valentine, R. W., SAE Automotive Engineering Congress, Paper No. 700156, Detroit, Mich., Jan. 12-16, 1970.
14. Rogers, A. E., Connolly, T. W., "Analog Computation in Engineering Design," McGraw-Hill, New York, 1960.

RECEIVED May 28, 1971.

13

Throwing Power as Related to Material Properties with Analysis by Digital Computer Simulation

A. E. GILCHRIST and D. O. SHUSTER

Glidden-Durkee, Division SCM Corp., Dwight P. Joyce Research Center, Strongsville, Ohio 44136

Electrodeposition is a process that can be defined by known electrochemical laws and inherent properties of the material deposited. Throwing power is that characteristic of the material being deposited that promotes uniform deposited film thickness, at varying cathode-to-anode distances on the same deposition electrode. Process variables such as deposition voltage, time, electrode areas and spacing, and tank size cannot be used as material properties. Electrical equivalent circuits and mathematical relations were developed to represent circuit actions derived. Such relations were combined as a digital computer program. Results of such computer calculations were tested by actual laboratory electrodeposition operations, and the results were compared. These comparisons show that electrodeposition throwing power is a characteristic of the deposition material and is not governed by application process variables.

Electrodeposition as a method of applying paint to a conductive object was patented by Davey (1) in 1919. Successful laboratory scale application of coatings on the inside of tinplate cans, after fabrication, was carried out in the 1930's (2). Commercial exploitation of this coating technique has only appeared, however, in the past 10 years (3). A comprehensive description of the process, paint compositions, and quantitative aspects of the electrical efficiency of the electrodeposition method of applying paint appeared in 1966 (4). Since then, additional details, such as the use of electrodialysis to control bath composition (5), resinous coating materials (6, 7), and the requirements for maintaining the bath composition in a continuous electrodeposition process (8) have been issued.

There is almost unanimous agreement that the deposition process obeys Faraday's law (4, 9, 10, 11)—i.e., the weight of deposit formed is proportional to the number of coulombs of electricity passed. Voltage (12), resin concentration, bath temperature, current density, and several substrates (10) do not influence the coulombic yield significantly. However, the coulombic yield does vary inversely with the degree of neutralization of the resin (10, 11, 13).

Since electrodeposition obeys Faraday's law, electrolysis—not electrophoresis—is the process governing the deposition. Ohm's law can be applied to the system to determine the amount of current flowing and voltage losses in the resistive elements comprising the system (9, 10, 12, 13). In contrast to electroplating, the specific conductivity of typical electrocoatings at operating solids and temperatures, 500–4000 micromho/cm, is so large that the bath resistance must be included in an analysis of the electrical circuit involved (9, 13). A more dramatic and significant difference from electroplating, however, is the electrical insulating property of the deposited film. It is the high specific resistance of the electrodeposited film which makes possible the uniform coverage of irregularly shaped objects and the coating of the inside of such restrictive areas as rocker panels on automobile bodies. This property of electrocoating is called throwing power although uniform thickness covering power would be a more accurate term for a pigmented paint film.

Much work has been done on the influence of the various reactions occurring at the anode and at the cathode on the electrodeposition process. In anodic deposition it has been demonstrated that 100% of the current flowing at the cathode can be accounted for by the electrolysis of water to generate hydrogen gas (14). Two competing reactions at the anode are proposed. The electrolysis of water at the anode will generate oxygen gas and hydrogen ions. Resin will precipitate on the anode as a result of the lower pH because of the increased concentration of the hydrogen ions. Alternately, it is proposed that metal ions generated by oxidation of the anode substrate or metal pretreatment react to precipitate the resin. Tawn and Berry (10), often falsely accused of supporting the role of metallic ions, actually present much data which support the acid coagulation theory. They demonstrated such evidence as the fact that the iron content of films on mild steel electrodeposited panels was no greater than on panels dipped into the same resins. The amount of iron found in the electrodeposited film was only a small fraction of that which would be required by the stoichiometry of the metal ion-precipitation mechanism or that required to account for the oxidation reactions occurring at the anode based on Faraday's law. Precipitated resin by addition of ferrous sulfate was insoluble in organic solvents such as methanol and acetone whereas the electrodeposited resin, which will also redissolve in the bath when deposition is terminated, was soluble in the organic solvents. Independently, it has been pointed out that the clear film obtained when an unpigmented resin is deposited on an iron anode is further evidence that the role of iron ions is inconsequential (15). It has also been observed that electrodeposition can be obtained on such inert surfaces as platinum and carbon (16). Recent studies of deposition over iron and platinum (17) demonstrate the dominance of the acid precipitation mechanism during the first two minutes or so of deposition, a time span most often found in commercial practice.

Wet on wet deposition of a white paint after a black, or vice-versa, was found to yield gray films (10). This example of the fact that the deposition layer builds up at the deposit–substrate interface rather than at the deposit–bath interface has been successfully demonstrated by a patented process (18). In this process the corrosion resistance of an electrocoated object is improved by immersing the wet coated object in a solution containing metal-treating oxyanions and electrodepositing them under the paint film. We have deposited red paint after black paint and vice-versa. Since the red pigment particles are much larger than the black, there is more surface discoloration of an initial black coating when red is forced through it on subsequent electrocoating than in the reverse situation. In either case, however, microscopic examination of a cross-section through the layers of coating clearly shows that the pigmented system which was deposited second is found between the pigmented material which is deposited first and the conducting substrate.

Experimental

Laboratory measurements of throwing power have been most noted for their variety of test apparatus (10, 13, 19). There is one example of a practical application of throwing power testing (20), a theoretical analysis (11), a summary of the relationships between material properties, deposition parameters, and throw (13), and a proposed relationship between throwing power and voltage, bath specific resistance and coulombic yield (11). Almost universally throwing power is recorded as inches of something. This number suffers because the basis for the measurement has not been defined, just as a specific resistance or specific conductivity as a number is meaningless unless the temperature at which the measurement was taken is also recorded. In order for a throwing power measurement to have any real significance, the film thickness developed on a reference anode must also be recorded (19).

Our purpose in this investigation was twofold. First, we wanted to identify the material properties of electrocoatings which influenced throwing power. Throwing power is a function or an index of deposited film thickness uniformity. It is not a function of total amount of film deposited nor the deposition conditions such as voltage and time.

Electrodeposition follows Faraday's law: one gram equivalent weight of matter is chemically altered at each electrode for 96,501 international coulombs of electricity passed through the electrolyte. The current passed can be derived from Ohm's law: voltage equals the product of amperes flowing and circuit resistance in ohms. When circuit resistance is constant, changes in voltage can only produce changes in amperes. When combined with time variations this produces only changes in total quantity of material deposited—not its distribution or location. The only factor influenced by the electrodeposition material is the circuit resistance, which is a combination of resistances of bath solution and deposited film material. If throwing power is an inherent property of a particular electrocoating formulation, such operating variables as voltage and deposition time are not parameters which determine throwing power.

Secondly, we wanted a mathematical description of the formation of a coating layer in a restrictive channel of definable geometry. If we could develop this mathematical description, we could use a computer to perform the throwing power experiment, based upon an input set of relevant material properties, physical dimensions of the test set up, and operating conditions. Success in step two of our objectives would enable us to develop a quantitative relationship between the relevant material properties and throwing power. The most important relevant properties are current required for deposition, (coulombs/gram), specific resistance of bath solution, and wet deposited film specific reistance. Differences between laboratory experiment and computer simulation could be analyzed to provide an even better understanding of the throwing power problem and thus lead to practical applications.

Apparatus

Electrocoating solution specific resistance measurements are performed in the manner established for electrolytes. The Yellow Springs Instrument Co. model 31 conductivity bridge, with an auxillary decade capacitor variable from 0.01 to 1.0 mfd has proved to be useful for this measurement. For routine use in the laboratory a conductivity cell was fabricated with polished stainless steel electrodes held rigidly in position in a machined block of Plexiglas. This cell is calibrated with KCl solutions in the standard manner. It is rugged, easily cleaned, and more efficient to use than platinized platinum electrodes which rapidly become surface contaminated with pigment, thus requiring frequent cleaning and replatinizing in order to maintain accuracy of measurement. With polished stainless steel electrodes 1000 Hz is mandatory to prevent polarization rather than 60 Hz commonly used with platinized electrodes.

Laboratory evaluations of the deposition properties, deposited film specific resistance and electrical efficiency (coulombs/gram or the reciprocal), are carried out as described previously (4). Current–time curves are recorded by a Sargent recorder model SR type or a Heathkit model EUW-20. Coulombs of electricity passed are measured either by a disc integrator mounted on the Sargent recorder or by a Self-Organizing System, Inc. model SI 100 electronic integrator. Net coating weight deposited is determined by weighing the 4-sq inch panels, before and after coating, on a laboratory analytical balance such as a Mettler type H5 (160 grams capacity).

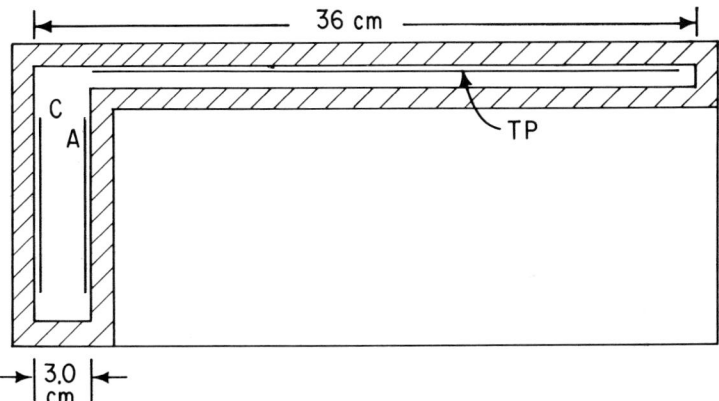

Figure 1. Plexiglas throwing power tank (top view; 0.4 scale). C, cathode; A, reference anode; TP, throwing power section. Base, 17 × 39 cm.

Laboratory throwing power experiments are performed with the test apparatus shown in Figure 1. The tank is 10 cm deep and is filled with 9 cm of bath. Four-sq inch (10 × 10 cm) panels are used for the cathode and the reference anode. In the throwing power section any surface up to a 4 × 12 inch (10 × 30 cm) panel may be used. When multiple sectional panels are placed in the throwing power slot, a sequence shunt switch is used to give current readings at discrete time intervals so that a current–time curve may be traced for each segment from the points recorded.

Results

Throwing power measurements are displayed by the method shown in Figure 2. Film thickness at several points along the throwing power

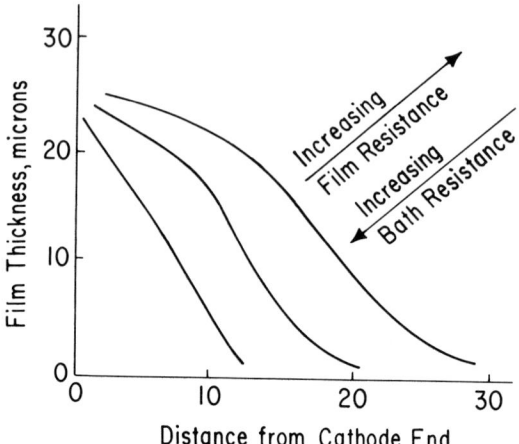

Figure 2. Method of displaying throwing power

panel is measured with a Permascope having the electronic circuit and measuring head appropriate for the type of substrate involved. Qualitative effects of deposited film specific resistance and bath specific resistance on throwing power are illustrated in Figure 2. The common denominator for the three examples shown is the same film thickness on the reference anode in each experiment. Film resistance can be varied essentially independently of other electrocoating properties by adding an organic solvent

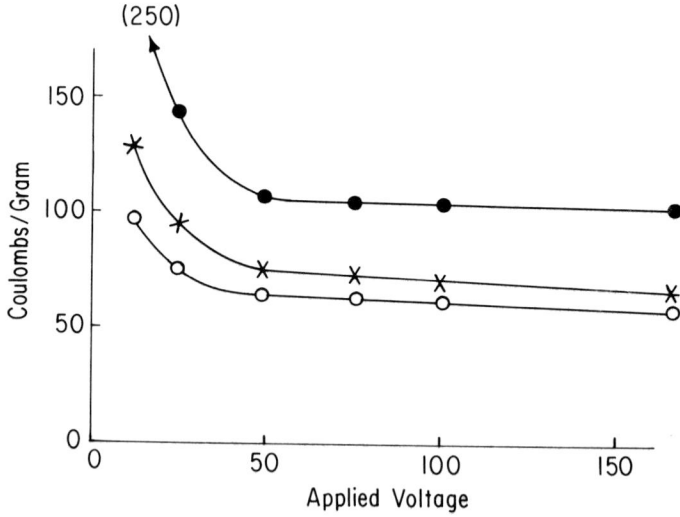

Figure 3. Coulombs/gram vs. voltage effect of level of pigment loading (composition by weight). ●, 100% vehicle; ×, 3 vehicle/1 pigment; ○, 2 vehicle/1 pigment.

to the vehicle before solubilization, thus altering the viscosity of the deposited film. Such viscosity changes of deposited film, whether by temperature or by solvent dilution, alter the specific resistance of the deposited material. Bath resistance can be varied independently of other properties by reducing the material to different solids levels.

The relationship between electrical efficiency and voltage is shown in Figure 3. Above *ca.* 50 volts, the coulombs/gram is essentially constant. Below 50 volts the curves climb toward undefined values as the voltage is reduced to the 5–10 volt range. A replot of the data in Figure 3 is shown in Figure 4. It is, however, still impossible to extrapolate the curves to a zero value of voltage and/or mg/coulomb.

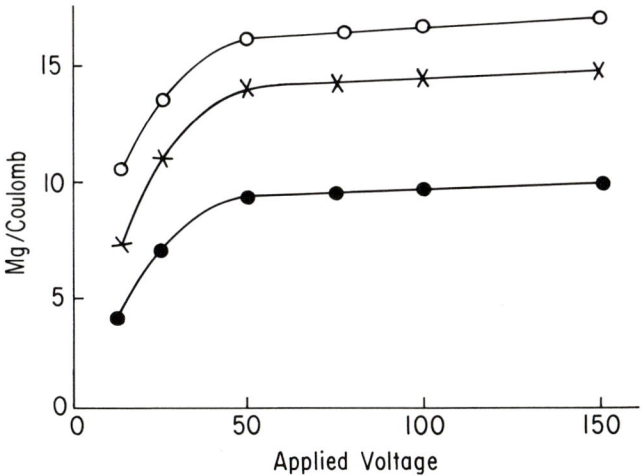

Figure 4. Milligrams/coulomb vs. voltage (replot of Figure 3)

To define the Faraday behavior of electrocoatings at low voltages, polarography experiments were performed. The current–voltage relationship in a typical polarography experiment is shown in Figure 5. The portion of this curve in the region from 0–5 volts is found in almost all physical chemistry texts in the first chapter on electrochemistry. This section of the curve shows the classic shape of the current–voltage curve for an electrolyte between inert electrodes (*21, 22*). The curve as shown in Figure 5 was taken directly from the strip chart recording trace of a programmed voltage dc power supply (selected linear increase of voltage with time) for a cell in which two cleaned cold rolled steel electrodes were inserted in a Plexiglas tank at an electrode separation of 5 cm. The surface area on each electrode exposed to the bath was 2.5 × 5 cm or 12.5 sq cm. As indicated, the slope of the straight line portion of the curve is the system resistance, which, for this experiment was calculated from the slope to be 62.5 ohms. Decomposition voltages in the range of 1.7–2.2 volts can be found in literature on electrolysis for electrolytes in which hydrogen is evolved at the cathode and oxygen at the anode.

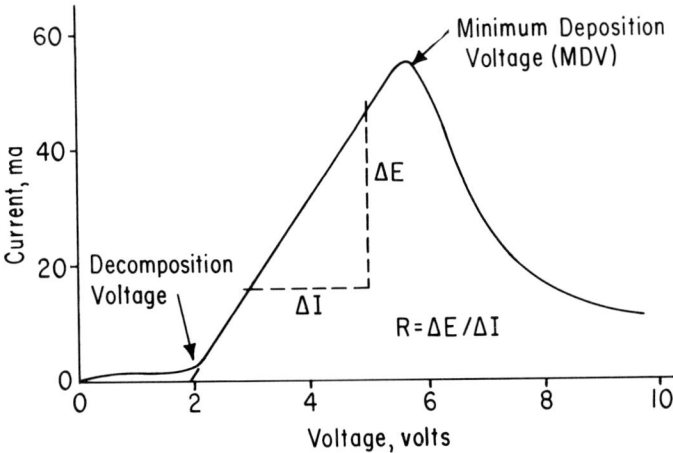

Figure 5. Decomposition voltage and minimum deposition voltage for an electrocoating material

There is apparently only one literature reference to a threshold deposition voltage of a paint (8) although minimum current densities are mentioned frequently. There is an example (16) in which, for two materials, (Figures 6 and 7 in Ref. 8) the flow of a finite amount of current (at 1.5 and 6 volts respectively) with a yield in each case of 0 grams of deposited dry film weight occurs.

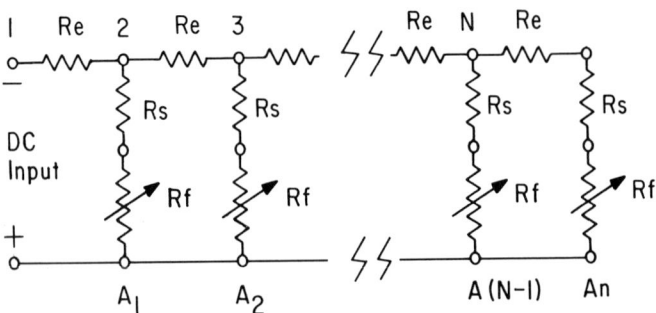

Figure 6. Electrical resistance circuit analyzed by a computer

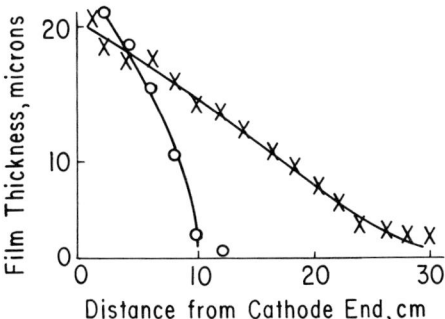

Figure 7. Comparison of experimental results with computer simulation. ×, *material A;* ○, *material B;* – –, *computer simulations.*

If the experiment as illustrated by Figure 5 is performed at different electrode spacings, the minimum deposition voltage (MDV) is a linear function of the distance between the electrodes, caused, of course, by the bath resistance. A plot of minimum deposition voltage *vs.* electrode separation extrapolated to 0 separation gives the true minimum deposition voltage. Since the relationship is linear, only two measurements, at 5 and 10 cm separation for example, are needed. By subtracting the difference between the MDV's at 5 and 10 cm from the value at 5 cm, the MDV at zero separation is obtained.

Digital computer simulation of the experimental measurement of throwing power was based on the following assumptions.

(1) A continuous, non-linear, time-dependent function can be treated as the summation of linear steps if the size of each differential time interval is small enough.

(2) The electrical resistance of an electrocoating bath confined by a given geometry can be calculated from the specific resistance of the bath at the temperature of the experiment, the dimensions of the chamber, and the area of the electrodes.

(3) The electrical resistance of the deposited film is a linear function of the film thickness and can be calculated by measuring the area of the electrode surface being coated and using the specific resistance of the deposited film in the way used to calculate the bath resistance.

(4) The current flowing to a given area of the electrode being coated at any time interval can be divided into that portion which causes deposition at the net coulombs/gram rate approached as a constant value (*see* Figure 3) and the remainder of the current which causes electrolysis but yields no deposition film.

(5) The experiment is performed under isothermal conditions or under conditions sufficiently close to isothermal that the rate of heat dissipation is much greater than the rate of heat generation and does not introduce a temperature effect on the film and bath resistance.

The equivalent electrical resistance circuit which the computer calculates is shown in Figure 6. For a given geometry throwing power experiment (*see* Figure 1, for example) the value for Re and Rs in Figure 6 can be calculated directly from the specific resistance of the electrocoating bath at the temperature and solids used for the experiment. For a cross-section of (1 cm × 9 cm) = 9 sq cm, Re is 1/9 of the specific resistance per centimeter of slot length. Since the distance from the center line of the slot to the anode surface is 0.5 cm, Rs is 1/2 of Re for a section of anode area 1 cm wide and 9 cm high. Rf is a variable resistance which is 0 for no deposited film and is a linear function of the film thickness.

The digital computer performs the following sequence of calculations for each interval of time selected and accumulates the results of successive buildups of deposited film on the anode segments A_1, A_2, etc.

(1) The resistance is calculated at the node points starting at point n and ending at point 1 which is the total resistance for the entire circuit during the time interval.

(2) By dividing this resistance into the applied voltage, the total current flowing through the circuit during the time interval is calculated.

(3) Voltage drop and current flow for each Re and $(Rs + Rf)$ are calculated.

(4) The portion of the current flowing through each element $(Rs + Rf)$ which results in deposition during the time interval is calculated by $(Vs - \text{MDV})/(Rs + Rf)$.

(5) Based on the laboratory measurements of the net coulombs/gram of deposit, specific resistance of the deposited film, area of the anode segment, and specific gravity of the deposited film, the amount of film thickness added during the interval and a new value for Rf based on the new film thickness are calculated.

(6) The computer repeats the cycle of calculations in steps 1–5 for a specific number of time intervals or until a selected film thickness is obtained on the first anode segment, A_1. The actual printout of information can be merely a table of the anode segment number and the film thickness developed on that segment after a given period of time. Alternately, printouts at selected time intervals of any of the desired voltages, currents, or other variables may be obtained.

Discussion

It was necessary to establish whether or not the model chosen for the digital computer simulation was a good one. A typical example of the kind of agreement between the laboratory experiments and the computer simulation is shown by Figure 7 for two materials having widely different throwing power. As Figure 7 shows, the agreement between theory and experiment is excellent. Obviously we have been able to identify the relevant material properties and their interrelationships as they relate to throwing power. Those properties are: specific resistance of the deposited film and specific resistance of the electrocoat bath at the temperature at the start of the experiment, decomposition voltage of the system for the electrodes employed, minimum deposition voltage (MDV) for the system, net electrical efficiency of deposition (coulombs/gram or mg/coulomb), and the specific gravity of the deposited coating. Visual evidence that voltage and time are not parameters which influence the throwing power of electrocoating is found in Figure 8. There is no argument that increased voltage or time does give more "inches of throw" for a given electrocoating; however a proportional increase in film thickness occurs on the reference anode as well. Therefore the throwing power was not changed.

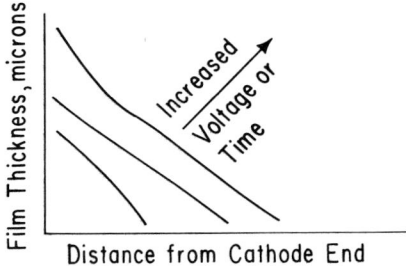

Figure 8. Effect of voltage or time on deposition

An example of the practical use of the digital computer simulation to evaluate the contribution of one of the relevant properties of electrocoatings to throwing power is shown in Figure 9. The curvature as indi-

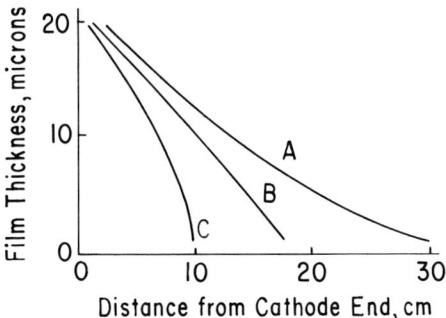

Figure 9. Effect of minimum deposition voltage on throwing power. Digital computer simulation for: MDV $A = 4$ volts, $B = 6$ volts, $C = 8$ volts.

cated in Figure 9 is computer predicted. We have found that the minimum deposition voltage of an electrocoating depends on the anode material. Qualitatively the MDV increases for steel substrates in the following manner: zinc phosphate (Bonderite 37), cleaned cold rolled steel, iron phosphate (Bonderite 1000), and untreated galvanized steel. Cleaned aluminum has a MDV higher than any of the above surfaces. Throwing power for a given electrocoating material was similar to that shown in Figure 9 for zinc phosphate, clean galvanized and untreated aluminum substrates in order of increasing minimum deposition voltages.

The most pronounced differences between laboratory experimental determinations of throwing power and computer simulation result from heat dissipation problems. It was possible to provide either no forced circulation or vigorous circulation in the slot of the throwing power tank (Figure 1) during deposition. In Figure 10 curve A is obtained with vigorous agitation and curve B with no agitation. To evaluate the heat dissipation problem further, a test strip was coated inside a pipe (20). When the cathode was placed at the bottom end of the tube, results

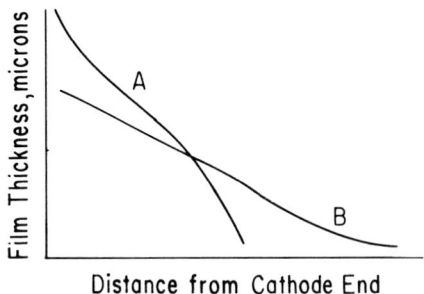

Figure 10. Effect of heat dissipation of film thickness distribution. A: cathode at top of tube or vigorous circulation in slot. B: cathode at bottom of tube or no circulation in slot.

similar to curve B in Figure 10 were obtained. When the throwing power tube was inverted so that the cathode was at the top end, results similar to curve A, Figure 10, were obtained. We have independently determined the film specific resistance and the bath specific resistance over a temperature range from 70° to 110°F. The temperature effect on bath specific resistance is slight over this temperature range; however the effect on the specific resistance of the deposited film is severe. Since the deposition occurs at the substrate–film interface and the heat is generated at this point and must be dissipated through the substrate and through the deposited film layer, the temperature effect on the specific resistance of the deposited film must be the dominant factor.

Summary

Electrodeposition is a reversible process, as are other simple electrolysis reactions in water solution. Deposition occurs at the electrode–deposited film interface if the applied voltage exceeds the minimum deposition voltage for the system. Operating conditions necessary to achieve commercially acceptable rates of coating application are far removed from equilibrium conditions. It is possible to simulate the actual deposition of an electrocoating in a cavity of defined geometry by an electrical circuit which may be analyzed by a digital computer. The major deviations between the model system and the actual experiment arise under conditions such that isothermal behavior is no longer valid. Although the energy input per unit time interval to each segment of the system is easily calculated, the rate and nature of the energy dissipation are much more complex problems.

Literature Cited

1. Davey, W. P., U. S. Patent **1,294,627** (Feb. 18, 1919).
2. Sumner, G. C., *Trans. Faraday Soc.* (1940) **36**, 272.
3. Phillips, G., "Electropainting for the Seventies," London, 1969.
4. Gilchrist, A. E., U. S. Patent **3,230,162** (Jan. 18, 1966).
5. Gilchrist, A. E., U. S. Patent **3,304,250** (Feb. 14, 1967).
6. Gilchrist, A. E., U. S. Patent **3,351,575** (Nov. 7, 1967).
7. Gilchrist, A. E., U. S. Patent **3,351,675** (Nov. 7, 1967).
8. Gilchrist, A. E., U. S. Patent **3,362,899** (Jan. 9, 1968).
9. Finn, S. R., Hasnip, J. A., *J. Oil Colour Chemists Assoc.* (1965) **48**, 1121.
10. Tawn, A. R. H., Berry, J. R., *J. Oil Colour Chemists Assoc.* (1965) **48**, 790.
11. Olsen, D. A., Boardman, P. J., *J. Electrochem. Soc.* (1967) **114**, 445.
12. Brewer, G. E. F., Hamilton, C. C., Horsh, M. E., *J. Paint Technol.* (1969) **41**, 114.
13. Hays, D. R., White, C. S., *J. Paint Technol.* (1969) **41**, 461.
14. LeBras, L. R., *J. Paint Technol.* (1966) **38**, 85.
15. Cooke, B. A., *J. Paint Technol.* (1970) **34**, 12.
16. Finn, S. R., Mell, C. C., *J. Oil Colour Chemists Assoc.* (1964) **47**, 219.
17. Giboz, J. P., Lahaye, J., *J. Paint Technol.* (1970) **42**, 371.
18. Gilchrist, A. E., U. S. Patent **3,290,235** (Dec. 6, 1966).
19. Brewer, G. E. F., Horsch, M. E., Madarasz, M. F., *J. Paint Technol.* (1966) **38**, 453.
20. Brewer, G. E. F., Strosberg, G. G., Horsch, M. E., *J. Paint Technol.* (1967) **39**, 551.
21. Maron, S. H., Prutton, C. F., "Principles of Physical Chemistry," pp. 573–574, MacMillan, New York, 1958.
22. Groening, A. A., Cady, H. P., *J. Phys. Chem.* (1926) **30**, 1597.

RECEIVED May 28, 1971.

Bath Maintenance and Design of Merchandise: Introduction

GEORGE E. F. BREWER

In conventional painting processes, the entire coating composition is transferred out of a pot, tank, or spray gun onto the article to be coated. The electropainting process, however, deposits only some of the total composition of the bath on the workpiece, leaving other elements in the bath. For example, the counterions in an electrocoat bath do not become part of the deposited film. Thus, the solubilizer stays in the bath while the resin is deposited and leaves the bath as a coat on the workpiece. Accumulation of solubilizer in the bath will eventually interfere with the coating process. Several methods have been developed to prevent solubilizer build-up. These methods are based on (a) removal of solubilizer left behind (Figure 1), or (b) re-use of surplus solubilizer to disperse resinous replenishment material (Figure 2).

The accumulation of solubilizer in the bath is only one of many disproportionations between bath and film composition which may occur and are known to occur. One phenomenon is the frequently observed tendency of inorganic pigments toward preferential deposition. For instance, a bath may contain 80 grams resin and 20 grams inorganic pigment per liter while the film deposited from the bath may contain 78 grams resin and 22 grams inorganic pigment. Preferential deposition of one resin, or resin variety, in the presence of others has also been observed (1, p. 64).

Actually, these disproportionations between bath and film composition pose analytical problems only since a bath will remain in its original composition if all components which leave are replaced in kind and quantity. Thus, the formulator of electrodeposition materials distinguishes between the composition of the bath and the composition of the feed. The composition of the bath is designed to produce test panels and sample pieces of the desired properties while analysis of the deposited film, the "dragged out" bath, and losses through evaporation, etc., determine the composition of the feed materials. A miniaturized coating installation can be used to evaluate feed composition (2).

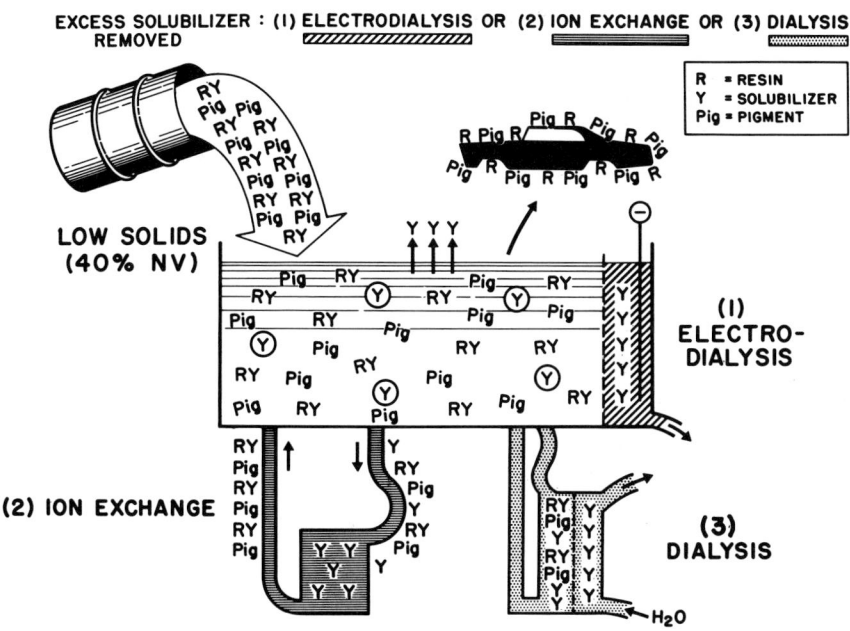

Figure 1. Solubilizer-satisfied feed

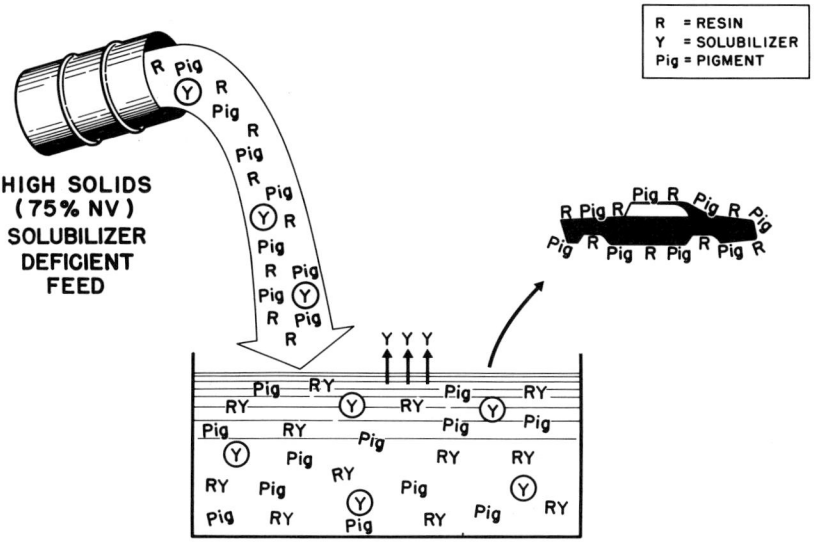

Figure 2. Solubilizer-deficient feed

Rinse and No-Rinse Process

Droplets remaining on the coated merchandise after it is lifted from the bath can adversely affect surface appearance. However, the interfacial surface tension between freshly deposited film and bath can be adjusted to ensure the free flow of bath droplets from the coated surface. Some installations are reported to use this "no-rinse" process. Highly formed workpieces may carry sizeable volumes of bath inside of cavities. To control appearance and film thickness on inner surfaces, most installations use a water or deionized water rinse.

For maximum recovery of the excess paint bath adhering to the workpiece, clear, resin-free bath fluid is separated from the tank by methods such as ultrafiltration (3). The ultrafiltrate contains some of the solubilizer and other truly dissolved substances, which make the ultrafiltrate the natural habitat of the paint colloids. The ultrafiltrate, therefore, prevents resin coagulation arising from infinite dilution. After serving as a rinse fluid, the ultrafiltrate is returned to the coating tank together with the rinsed-off bath. In other words, the paint bath carried inside of cavities is replaced by an ultrafiltrate of low paint solids content.

Literature Cited

1. Brushwell, W., "A Review of Electrocoating," *Am. Paint J.* (July 13, 1970) 71–77 and (July 27, 1970) 64–70.
2. Burnside, G. L., Strosberg, G. G., Brewer, G. E. F., "Electrocoating Process: Bath Maintenance and Engineering Principles," *Paint Varnish Prod.* (1965) **55** (12), 53–66.
3. Bardin, P. C., "A Departure in Electropainting," *Ind. Finishing* (1970) **46** (8), 26–30, 32.

14

Material Balance Considerations in an Electrocoating Tank

WILLIAM VAN HOEVEN, JAMES E. LOHR, and
WILLEM B. VAN DER LINDE

Marshall Research and Development Laboratory, F & F Department, E. I. du Pont de Nemours & Co., Inc., 3500 Grays Ferry Ave., Philadelphia, Pa. 19146

A mathematical model has been developed which describes the time-dependent concentration behavior of component(s) in an operational electrocoating bath. The equation is:

$$C_T = C_o e^{-k_1 T} + k_2 \sum_{n=0}^{T-1} e^{-k_1 n}$$

where C_T is the component concentration at time, T, C_o is the initial concentration, k_1 is the time-dependent first-order rate constant for component removal, and k_2 is the linearly time-dependent constant for component addition. The model is based upon the addition being a stepwise process and removal being a continuous process. The validity and scope of the model are demonstrated in laboratory experiments and field situations. Examples include application to the concentrations of solvent, amine, and water-soluble nonvolatiles. The value of such information for assistance in formulating, controlling, and designing electrodeposition systems is shown.

It is a critical prerequisite that the material balance in an electrocoating tank be kept within set values to ensure satisfactory performance at any time during the life of the bath. In other words, the concentration of all components of the paint, including contaminants, must be maintained within certain limits to permit the deposition of acceptable coatings. In view of the large number of components which go into most electrocoating formulations, this is a complex problem. Historically the attention given this problem is as old as electrodeposition itself. Forty-year old patents and journal articles, concerned with rubber electrodeposition, consider bath composition and material balance (*1*). Recent literature contains numerous articles discussing differential deposition of pigments (*2*), accumulation of solvents (*3*), and the difficulties of maintaining the proper amount of amine solubilizer (*4*).

Paint deposition occurs when dispersed particles are electrochemically destabilized in the region of the substrate. In general, the total system is very sensitive to electrochemical changes. Since the components and contaminants in an electrocoating tank contribute in different ways to the electrochemistry and the stability of the dispersion, it is difficult to quantify effects without extensive analytical efforts.

Nevertheless, commercial electrodeposition is a reality despite the delicate nature of such systems. This has been made possible because the addition and removal of materials can be systematically and accurately controlled. The purpose of this paper is to show how the systematic changes occurring in an electrocoating tank can be calculated and used to predict and to control the concentrations of the components in the tank.

To maintain a given level of any component in a tank, it is necessary to know, quantitatively, the modes of addition and removal of that particular species. This is possible only with a thorough understanding of the mechanism of electrodeposition and the distribution of the components between the continuous and discontinuous phases in the electrocoating bath. The rates of addition and removal of each component of the electrodeposition bath must also be known. Fortunately many of these input and removal rates are restricted within fixed limits by the equipment design, turnover rates, paint formulas, etc. Furthermore the treatment is simplified by the fact that electrocoating tanks are maintained at constant volume.

The proposed material balance model takes into account the rates of component addition and removal, by whatever means, and allows the calculation of the concentration of that component at any given time.

Removal of bath components involves several factors. Certain species —*e.g.*, solvents, crosslinkers, etc.—may partition between the organic and aqueous phases so that they will be partially removed by deposition of paint (*3, 5*). If components do not partition into the aqueous phase but remain associated with the organic binder, they are removed in direct proportion to the rate of depletion of paint solids.

If an electrodeposition tank is equipped with an ultrafiltration unit, or some other device which selectively removes the aqueous phase, species dissolved in this phase will be removed at a rate which depends on the instantaneous concentrations in the bath—*i.e.*, the rate of removal of such species is first order with respect to their concentration. That is:

$$\frac{dc}{dt} = -k_1 c \qquad (1)$$

Integration over the limits $t = 0$ and $t = T$ gives:

$$C_t = C_o e^{-k_1 t} \tag{2}$$

Where C_t is the concentration at time T, C_o is the concentration at $t = 0$, and k_1 is the slope of a plot of ln C vs. T.

Addition of materials to the bath is more straightforward. For any species which enters the paint bath as part of the replenishment, or with the work, there is a rate of increase in concentration, k_2, which is time dependent only. That is:

$$C_t = C_o + k_2 (T - T_o) \tag{3}$$

or:

$$k_2 = \frac{\Delta C}{\Delta T} \tag{4}$$

where ΔC is the amount of that component brought into the tank in ΔT time units. For each operation, ΔT is the number of days required for one turnover—i.e., the number of days required to remove an amount of paint solids equal to the amount originally present in the tank. Of course if there is no removal mechanism, the species continues to increase in concentration. At low concentrations, ionic contaminants approximate this behavior.

In an operating system, the addition and removal expressions can be combined to obtain concentration at any time $T > T_o$. In the combined expression it is assumed that T is in days and that the tank is operated such that an amount of the species sufficient to produce a concentration change of magnitude k_2 ($= \Delta C/\Delta T$) is added at the end of each operating day. Thus at the end of the first day's operation, following the replenishment addition, the concentration of the component in question will be $C_1 = C_o e^{-k_1} + k_2$; at the end of the second day, $C_2 = (C_o e^{-k_1} + k_2)e^{-k_1} + k_2$; etc. At any time T,

$$C_T = C_o e^{-k_1 T} + k_2 \sum_{n=0}^{T-1} e^{-k_1 n} \tag{5}$$

which can be modified to:

$$C_T = C_o e^{-k_1 T} + \frac{k_2 e^{k_1}}{e^{k_1} - 1} \left(1 - \frac{1}{e^{k_1 T}}\right) \tag{5a}$$

Using Equation 5,

$$\text{as } T \to \infty$$
$$C_\infty \to \frac{k_2}{1 - e^{-k_1}} \qquad (6)$$

when the time for incremental additions equals one day.

In the ideal case, where addition and removal are simultaneously continuous, the concentration change with time is:

$$\frac{dc}{dt} = k_2 - k_1 c(t)$$

Integration of this function gives:

$$C_t = C_o e^{-k_1 t} + \frac{k_2}{k_1}(1 - e^{-k_1 t}) \qquad (5')$$

It is evident that in this case the asymptotic value of C_t is:

$$C_\infty = \frac{k_2}{k_1} \qquad (6')$$

Since $1 - e^{-k_1} \approx k_1$, Equation 6' can be used in process design to determine k_1 for given values of C_∞ and k_2.

To simplify the use of rate Equation 5, a Hewlett-Packard 9100 calculator/plotter system has been programmed to give C vs. T plots directly, for preselected k values and initial conditions.

The rate equation in the "step-wise" form is especially useful for simulating actual operating situations. By using a summation instead of an integral solution, compensation for unexpected random additions of any species can be made. Also, variations in rates of removal can be accommodated. If a component is being removed by more than one first-order concentration dependent mechanism, k_1 can be replaced by a composite k_1 which is simply the sum of individual first-order k_1's. In practice a composite k_1 is often necessary, as in the case of solvents, where deposition with the coating, evaporation, and ultrafiltration must all be considered. We have found experimentally that when electrocoating formulations containing less than 15% solvent are held in controlled environments, plots of ln C (solvent) vs. T are essentially linear; the slope of such a plot is the k_1 caused by evaporation. These experimentally derived evaporation rate constants are useful components of the composite k_1's used in subsequent concentration vs. time calculations.

Figure 1 demonstrates the utilization of the model. In this example, arbitrary but realistic values of C_o, k_1, and k_2 have been used. Curve A

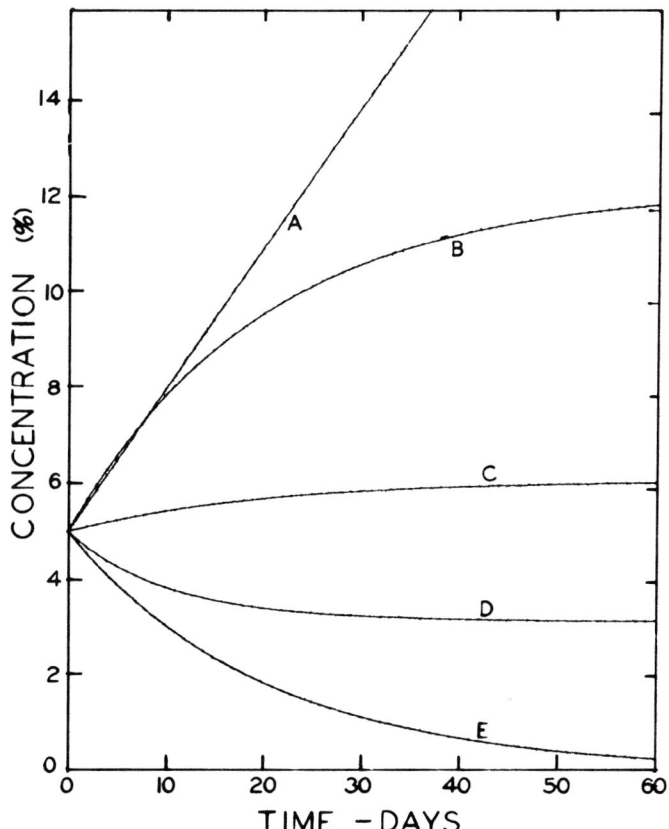

Figure 1. Concentration vs. time curves according to Equation 5

Curve	C_o, %	k_1, % day^{-1}	k_2, % day^{-1}
A	5	0	0.3
B	5	0.05	0.3
C	5	0.05	0.3
D	5	0.1	0.6
E	5	0.05	0

represents the increase in component concentration when no removal mechanism is operating (*i.e.*, $C_o = 5\%$, $k_1 = 0$, and $k_2 = 0.3\%$ day^{-1}). Curve E represents the opposite extreme of removal without addition ($C_o = 5\%$, $k_1 = 0.05$ day^{-1}, $k_2 = 0$). Curve C is the result of combining these two extreme situations and represents the concentration *vs.* time behavior of any species which is subject to these selected addition and removal rates and initial concentration ($C_o = 5\%$, $k_1 = 0.05$ day^{-1}, $k_2 = 0.3\%$ day^{-1}). Curve B shows the change which occurs when C_o and k_1 are held constant, and k_2 is doubled ($C_o = 5\%$, $k_1 = 0.05$ day^{-1}, $k_2 = 0.6\%$ day^{-1}). This is equivalent to doubling the production rate, going to a two-shift operation, etc. Curve D shows the effect of doubling k_1 while holding C_o and k_2 at the original value ($C_o = 5\%$, $k_1 = 0.1$ day^{-1}, $k_2 = 0.3\%$ day^{-1}). This situation would arise if the evaporation rate were doubled or if the ultrafiltration rate were increased.

Although the rate at which a component approaches its equilibrium concentration, C_∞, is often important, the value of the equilibrium concentration is generally of even greater concern. If, for example, Curve B of Figure 1 represents a component concentration behavior for a given paint formulation, the equilibrium concentration can be calculated to be 12.3%. If this level of the component adversely affects the electrocoating process, the model can be used to calculate the value of k_1 necessary to maintain any tolerable concentration of the species. For example, if the equilibrium concentration of Curve C is an acceptable value, the paint must be reformulated with a different but similar component with twice the evaporation rate, or the tank design may be altered to achieve the same goal, or ultrafiltration may be increased. Thus the model can be used to predict future difficulties and, by knowing the magnitude of the problems, assist in preparations to correct these difficulties.

Perhaps the biggest problem in electrodeposition has been controlling the concentration of the amines used to solubilize the resin. Approaches have involved the use of volatile amines, amine deficient replenishments, and flushed cathodes. The mathematical model discussed here is directly applicable to the first two approaches. If flushed cathodes are used, this model does not apply since the rate of removal of amine does not depend on its concentration but only upon the amount of current used for coating. The use of volatile amines is exactly analogous to the previous discussion where the model was shown to be very useful.

Motoyama *et al.* have published results of the amine balance in a system utilizing amine deficient replenishment (6). By using a non-volatile amine, evaporation is disregarded, and removal is limited to the amount carried out on the work piece in the combined film (deposited film plus drag out). Since this amine inclusion has a first-order concentration dependence, it is amenable to treatment by our model.

In this system, the initial amine concentration, C_o, is 8.6 meq/20 grams solids. Removal of amine by the electrocoating process is a linear function of amine concentration in the tank. The ratio of the amine concentration in the combined film to amine concentration in the bath is 0.34. The average turnover rate is 2.3 turnovers per month (one turnover per 13.26 days). Therefore, with no amine addition, using Equation 1,

$$\frac{dc}{dt} = \frac{0.34}{13.26 \text{ days}} c(t) \quad (t \text{ in days})$$

or, $k_1 = 0.0256$ day^{-1}.

The amount of amine added in one turnover increases the concentration 5.5 meq/20 grams solids if no amine is removed. That is:

$$k_2 = \frac{5.5 \text{ meq}/20 \text{ grams solids}}{13.26 \text{ days}}$$

or

$$k_2 = 0.4135 \text{ (meq}/20 \text{ grams solids) day}^{-1}$$

Using these values of C_o, k_1, and k_2, the change in amine concentration in the bath with time can be calculated. Figure 2 compares the computed concentration vs. time curve with Motoyama's data. Using Equation 6, $C_\infty = 16.35$ meq/20 grams solids, which compares favorably with 16.18 meq/20 grams solids calculated by Motoyama.

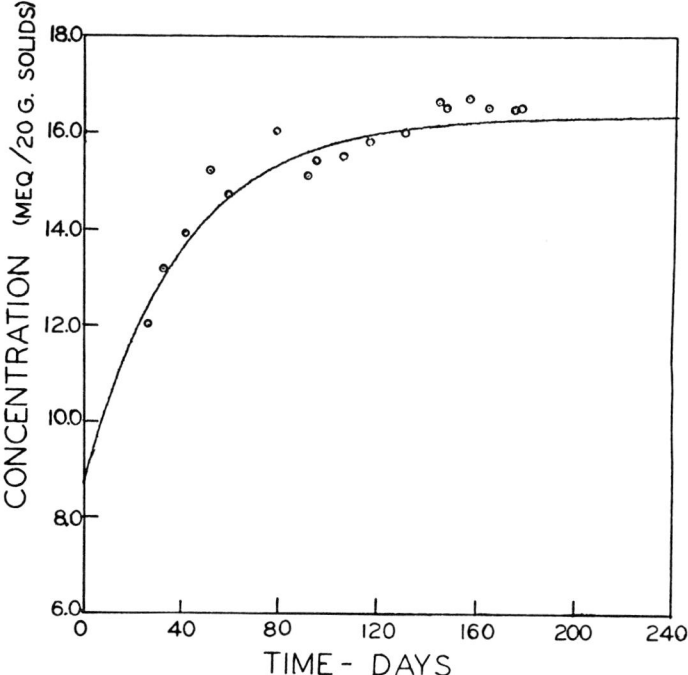

Figure 2. Comparison of amine concentration vs. time according to the data of Motoyama (6) and calculation using Equation 5. For the curve, $C_o = 8.6$ meq/20 grams solids, $k_1 = 0.0256$ day^{-1}, $k_2 = 0.4135$ (meq/20 grams solids) day^{-1}.

In commercial electrocoating systems there are often a considerable number of operational inconsistencies, on a day to day basis, in the addition and removal of materials. Daily additions of replenishment compo-

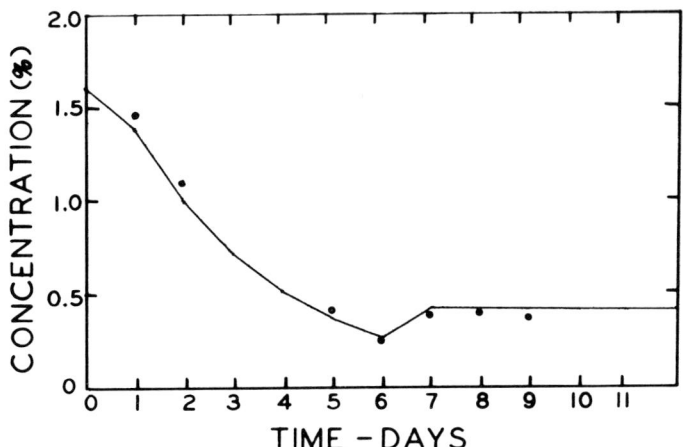

Figure 3. Comparison of data and calculated component concentration vs. time when addition rate, k_2 varies. $C_o = 1.60\%$, $k_1 = 0.331$ day^{-1} (constant throughout). Initial $k_2 = 0.231\%$ day^{-1}; after day 1, $k_2 = 0.0\%$ day^{-1}; after day 6, $k_2 = 0.231\%$ day^{-1}; after day 7, $k_2 = 0.112\%$ day^{-1}.

nents may vary, as does the amount of coating. As mentioned previously, random events can be taken into account when using this model. Figure 3 is a plot of component concentration vs. time for such a situation. C_o was determined analytically to be 1.60%. In this case, k_1 remained constant (0.331 day^{-1}) through the entire time interval of 10 days. The value of k_2 varied four times during this time period, ranging from zero to 0.231% day^{-1}, depending on the amount of replenishment added. The points are actual data based on analytical determinations of the component concentration. The agreement of the calculated and analytically determined values demonstrates that the application of this model to commercial operations is practical.

The model described in this paper is an organized, mathematical approach to the material balance of an electrocoating tank. The approach recognizes the stepwise modes of addition and the first-order nature of depletions and incorporates both into a single equation which describes the instantaneous and equilibrium concentration of any component in the electrocoating bath. The model can be applied to both paint formulation and process design.

Acknowledgments

The authors thank R. E. Wheeler for his assistance with the mathematics, L. F. Nonemaker for his guidance and assistance in the preparation of this manuscript, and E. I. du Pont de Nemours & Co., Inc. for permission to publish this paper.

Literature Cited

1. Sheppard, S. E., *Trans. Electrochem. Soc.* (1927) **52,** 47.
2. Robinson, F. D., Tear, B. J., *J. Oil Col. Chem. Assoc.* (1970) **53,** 265.
3. Tsou, I. H., U. S. Patent **3,434,952** (1969).
4. Burnside, G. L., Brewer, G. E. F., Strosberg, G. G., *J. Paint Technol.* (1969) **41** (534) 431.
5. Koral, J. N., Blank, W. J., Falzone, J. P., *J. Paint Technol.* (1968) **40** (519) 156.
6. Motoyama, Y., Kusano, H., Ohe, O., *J. Paint Technol.* (1969) **41** (533) 402.

RECEIVED May 28, 1971.

15

Turnover Studies on Amino Crosslinked Electrocoating Paints

WERNER J. BLANK

American Cyanamid Co., Industrial Chemicals and Plastics Division, 1937 West Main St., Stamford, Conn. 06904

Radioactive tracer work was found to be a useful tool in determining the amine balance in an electrocoating tank. The amine which codeposited with the polymer, the amine which was removed from the bath with the drag out liquid, and the amine content of the feed composition were found to determine the steady state of an electrocoating tank. Using the data obtained in the radioactive tracer work it was possible to predict accurately the rise in amine content from a 35% neutralized bath to a steady-state level of 65% neutralization. The ability to predict the amine equilibrium will simplify future turnover studies and will decrease the amount of work required to select an amine for a certain resin and paint formulation.

The use of amino crosslinked electrocoating paints has substantially increased during the last few years. Their performance makes them especially suitable for white and pastel colors and for applications requiring increased outdoor durability and high detergent resistance. Their advantage over self-crosslinked polymer systems lies in the ease with which their crosslinking density can be adjusted according to the requirements.

The paint studied in the present work was a high reflectivity white formulation based on an acrylic resin, crosslinked with an amino crosslinking agent. This formulation gives the minimum required reflectivity of 85% for lighting fixtures. In our previous work (1) we demonstrated the performance and static stability of such a system. In this paper we show the consideration for achieving turnover stability.

Experimental

Formulation. The formulation used in this study was developed for white applications requiring high reflectivity at about 1 mil film thickness. This demanded high pigment-to-binder ratio in the electrocoating bath. The ratio used here was 60/100. As backbone resin we selected an acrylic polymer with an acid number of approximately 100. This material is available under the commercial name of XC-4010 (American Cyanamid Co. Industrial Chemicals and Plastics Div., Wayne, N. J.). The crosslinking agent selected was a hexa(alkoxymethyl)melamine which is available under the commercial name of Cross-Linking Agent 1116 (CLA-1116) (American Cyanamid Co., Wayne, N. J.). The resin-to-crosslinking agent ratio used was 72/28.

The pigment was an aluminum oxide-treated titanium dioxide rutile pigment (Unitane OR-600 titanium dioxide) (American Cyanamid Co. Pigments Div., Bound Brook, N. J.).

The neutralizing amine used was diisopropanolamine. It was selected because it allows the preparation of stable feed materials at low neutralizations. The initial neutralization of the bath was 35%—i.e., 35% of all available carboxyl groups of the resin were neutralized with amine. Bath and feed formulation are shown in Table I.

Table I. Bath and Feed Formulation

	Bath	Feed
XC-4010 resin	66.7	61.5
Crosslinking agent 1116	12.5	9.0
Diisopropanolamine	4.3	2.3
Unitane OR-600 titanium dioxide	37.5	46.0
Deionized water	79.0	—
Bath liquid	—	81.2

The electrocoating paint was prepared by grinding the XC-4010 resin, crosslinking agent 1116, diisopropanolamine, and titanium dioxide blend on a three-roll mill and dispersing the resulting viscous paste on a high speed dissolver. The water must be added to the paste very slowly to ensure uniform dispersion.

The composition of the feed material was based on the consideration that the reduction in bath solids takes place by deposition and by drag out (mechanical entrainment) of the bath liquid. To obtain a steady-state condition, the feed material must have the weighted average composition of the deposited film and the drag out liquid. By analyzing the deposited film and the drag out liquid as described below, we arrived at a feed composition with a pigment-to-binder ratio of 85/100 and a resin-to-crosslinking agent ratio of 76.5/23.5. The neutralization of 20% was the lowest possible since at lower levels the emulsified paint had a very large particle size and was unstable. The coating bath referred to as bath liquid was used to emulsify the feed material. The amine content in the coating bath enhanced the emulsification of high solids feed.

Analysis of the Deposited Film. Attenuated total reflectance spectroscopy (2) (ATR) was used to analyze the ester-to-triazine ratio in the deposited film. The pigment content of the deposited film was determined by solvent washing of the deposited uncured paint into a crucible. After evaporation of the solvent at 105°C for 2 hours the resin and crosslinking agent were burned off, and the inorganic residue was weighed.

Determination of the Amine Content by a Radioactive Tracer Method. PREPARATION OF THE RADIOACTIVE AMINE. Tritiated diisopropanolamine (DIPA) was prepared by exposing 0.73 gram of finely ground inactive DIPA to 4.5 curies of pure tritium gas for 27 days. After exposure, the radioactive DIPA was degassed several times to remove any excess tritium gas. To remove any water-exchangeable tritium, the amine was dissolved in 15 ml of water, filtered, and evaporated to dryness. This process was repeated until the specific activity of the DIPA-H (3) no longer changed. Only the tritium atoms attached to the N atom and O of DIPA can be exchanged with the hydrogens of the water. By evaporation of the water the exchangeable tritium is removed as a component of the water, leaving the stable tritium–carbon bonds unaffected.

After this repeated evaporation process, 0.5751 gram of DIPA was recovered and dissolved in 10 ml of deionized water, then .05 ml of this solution was diluted with deionized water to 100 ml. An aliquot containing 2.875 μgrams DIPA was dissolved in a modified Bray (3) solution and counted, giving an observed activity of 38,950 counts/min. With a counting efficiency of 29%, this activity corresponded to 21 millicuries/gram or 46,700 absolute disintegrations per minute per μgram of DIPA. A Packard M 3375 liquid scintillation spectrometer was used for the counting. For a convenient and precise handling, the active amine was blended with 1.9250 grams of inactive amine and diluted to 25 ml with deionized water.

ADDITION OF ACTIVE AMINE TO THE COATING BATH. Five ml of the above-described amine solution was used in combination with inactive amine to prepare 1 liter of a 10% electrocoating bath. The formulation of this bath was the same as described for the formulation of the tank, except that the neutralization was 20%. The neutralization was raised during the experiment by adding inactive amine to 40, 60, 80, and 100% neutralization.

SPECIFIC ACTIVITY IN THE DEPOSITED ENAMEL. The uncured electrodeposited enamel on the panel was analyzed for amine content by dissolving a known amount of electrodeposited paint in 20 ml of Bray solution and analyzing aliquots of 0.4–4 ml as described above.

Bath Analysis during Turnover. SOLIDS DETERMINATION. A 10-ml sample (accurately measured) was taken from the well-stirred bath and transferred into a weighed aluminum dish of 9.5 cm diameter, then dried in an oven for 20 minutes at 175°C.

pH MEASUREMENTS. The pH was measured with an electronic pH meter with glass electrodes with an accuracy of ±0.05-pH units.

CONDUCTIVITY MEASUREMENTS. A Wheatstone bridge with a frequency of 1000 Hz was used. The platinum black-coated electrodes of the conductivity cell gave a measuring range from 0.1 to 2000 micromhos/cm. The measurements were made at 25°C ± 1°C.

RESIN-TO-CROSSLINKING AGENT RATIO. The same method (2) described in the film analysis was used.

PIGMENT-TO-BINDER RATIO. Ten ml of bath liquid were pipetted into a tared crucible and dried for 2 hours at 125°C. The crucible was weighed, the binder burned off in a furnace, and the crucible was reweighed.

AMINE TITRATION. Ten ml bath liquid were diluted in a 250-ml beaker with 50 ml of deionized water; 2 grams of Triton X-100 surfactant (Rohm & Haas Co.) were added, and the blend was agitated with a magnetic stirrer for 5 minutes to obtain complete dispersion. The blend was titrated with 1-ml increments of 0.1N HCl in the conventional manner to a final pH of about 3. From a plot of pH $vs.$ ml HCl, the volume of HCl used to reach pH of 4.6 was determined.

COULOMBIC YIELD. The coulombic yield of the paint was determined by deposition at a constant voltage on an untreated steel panel. During the deposition process the curve of current $vs.$ time was integrated manually or with a Lectrocount II integrator (Royson Engineering Co., Hatboro, Pa.). The panels were reweighed after baking. The coulombic yield is expressed in mg/coulomb.

Method of Turnover Determination. CONTINUOUS COATING APPARATUS. For this study we used an Abrex (Abrex Specialty Coatings Ltd., 280 Wyecroft Rd., Oakville, Ontario, Canada) depletor which utilizes a rotating steel wheel as the anode. The uncured paint is continuously scraped off the wheel. This machine permits a high turnover rate and the determination of the feed composition. It does not simulate the introduction of any impurities ($e.g.$, from metal pretreatment) under production conditions. The initial turnover rate was half a turnover per day, and it was later increased to one turnover per day. At the end of each turnover, checks of all possible parameters were made using the techniques described above. The wheel was run at about 1 rpm, which gives a drag out of approximately 15%. The deposition voltage was 130–140 volts with an immersion time of 20–25 seconds.

FEED ADDITION. During the turnover the solids fluctuated between 10 and 8%. When the solids level fell to 8%, it was increased to 10% by adding a highly concentrated feed material. The amine deficient feed was emulsified on a high speed dissolver by the slow addition of bath liquid. The composition of the feed material remained unchanged during the entire series of turnovers.

Results

Film Composition. RESIN-TO-CROSSLINKING AGENT RATIO. Based on the infrared (IR) analysis of the deposited film, a feed material containing a resin-crosslinking agent ratio of 76.5/23.5 was used. This corresponds to approximately 85% migration efficiency and codeposition of the amino crosslinking agent—$i.e.$, the proportion of crosslinking agent in the deposited binder was 85% of its proportion in the binder content of the bath.

PIGMENT-TO-BINDER RATIO. The pigment-to-binder ratio of the deposited film, by analysis, was 90/100. Assuming a drag out of about 15% (found to be substantially the value in actual operation) and the drag out to have the same composition as the bath, we arrived at a pigment-to-binder ratio of 85/100 for the feed.

AMINE CODEPOSITION. The amount of amine codeposited with the paint at the anode has a significant influence on the amine balance during a turnover. To obtain steady state, the amount of amine added during the turnover must be the same as the amount lost by evaporation, drag out, and codeposition. Amine losses through evaporation were found to be negligible for this system.

The necessary amine content of the feed material depends on the hydrophobic–hydrophilic balance of the binder system and the stability requirements of the feed. Figure 1 shows that the percentage of the amine in the deposited film rises with the deposition voltage. The or-

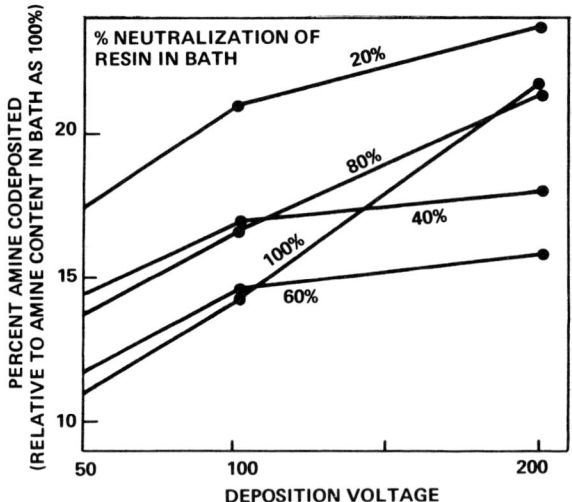

Figure 1. Radioactive tracer studies on high reflectivity paint. Resin, XC-4010; crosslinking agent, CLA-1116; pigment, Unitane OR-600 titanium dioxide; amine, tritium-labeled diisopropanolamine.

dinate in these graphs gives the percentage of amine in the film relative to the bath content as 100%. Thus if the bath resin was 40% neutralized, a 17% amine content in the film corresponds to 6.8% neutralization in the film. Figure 1 also demonstrates that the percentage of amine codeposited decreases with higher neutralization and reaches a minimum at about 60% neutralization. In the previous work (*1*) we have shown that this level of neutralization provides the optimum in throwing power (*4*). The percentage of amine codeposited at neutralization higher than 60% and higher voltages increases, thus reducing the throwing power.

In Figure 2 the percentage of amine codeposited is shown as a function of the deposition time. The amine concentration does not change

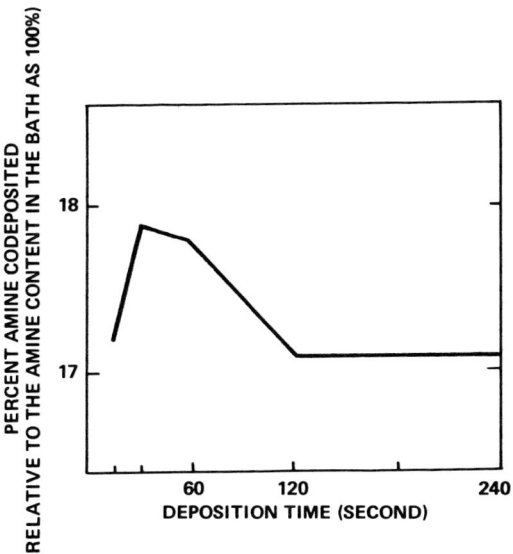

Figure 2. Radioactive tracer study for high reflectivity paint. Resin, XC-4010; crosslinking agent, CLA-1116; pigment, Unitane OR-600 titanium dioxide; amine, tritium-labeled diisopropanolamine; neutralization, 40%; deposition voltage, 200 volts.

with the deposition time, which is surprising considering the high voltage drop across the film and the increase in film resistance with time.

Turnover Results. These results are summarized in Table II.

RESIN-TO-CROSSLINKING AGENT BALANCE. The initial resin-to-crosslinking agent ratio of 72/28 is kept essentially constant by the feed material containing 76.5/23.5 resin-to-crosslinking agent ratio. There is a slight increase of about 1–2% in the amino crosslinking agent, which has no substantial effect on the properties of the paint.

PIGMENT-TO-BINDER BALANCE. The initial pigment-to-binder ratio of 60/100 increases slightly to 65/100 utilizing a feed material with a pigment-to-binder ratio of 85/100. A feed with a pigment-to-binder ratio of 80/100 would have been a better choice to maintain exactly the 60/100 ratio in the bath.

AMINE BALANCE. At the start-up, the bath had been low in neutralization. Only 35% of all available carboxyl groups were neutralized with diisopropanolamine. The feed material had the lowest possible neutralization for adequate dispersion stability—namely 20%. As shown in Figure 3, the amine level (meq) rises slowly up to the sixth turnover and then remains constant up to the tenth turnover. The increase in amine level corresponds to a neutralization to 65%. At this level the performance of

Table II. Turnover Study

Property [a]	Turnover Number			
	0	1.18	1.96	3.05
1. pH	8.0	8.0	7.9	7.9
2. Conductivity	495	545	595	624
3. Percent Solids	10.0	10.9	9.9	10.2
4. Coulombic yield	65.6	68.0	53.2	48.8
5. Film weight	6.2	6.0	5.6	6.6
6. Throw ratio	800	870	690	505
7. Pigment binder	60/100	71/100	65/100	62/100
8. Resin/crosslinker	72/28	76/24	65/35	72/28
9. Meq/100 grams XC-4010	68.3	76.5	107	98
10. Meq/100 grams Solids	35.6	39.4	47.1	48.4
11. Dragout	0	16.6	15.8	16.6
12. Reflectivity	88/88	88/88	89/88	87/87
13. Gloss 60°	68/74	70/72	67/70	69/76
14. Knoop hardness	5.6	5.7	7.9	6.6

[a] Properties which are not self-explanatory:
  2. Conductivity: in μohm^{-1} cm^{-1}.
  4. Coulombic yield: mg of paint deposited/coulomb.
  5. Film weight: mg of paint deposited/cm^2.
  6. Throw ratio: measured in a special cell with constant distance of cathode and anode at constant voltage for 1 minute.

$$\text{Throw Ratio} = \frac{\text{initial current} \times \text{voltage}}{\text{current after 1 minute}}$$

the bath is still good, and no changes in film properties are observed. Utilizing a two-component feed, which allows lower amine level in the feed material, the bath could have been run at lower neutralization.

Assuming that loss of amine occurs by codeposition at the anode and by drag out of bath liquid—but not by evaporation—and that losses are compensated through addition of fresh amine with the feed material, we can use following equation to calculate the required amine level in the deposited film to achieve steady-state conditions.

$$A_1 = \frac{A_3 - D A_2}{(1 - D)}$$

A_1 = amine codeposited with the paint (to be calculated)
A_2 = amine in the drag out (bath liquid)
A_3 = amine in the feed material
D = drag out

on a High Reflectivity Paint

			Turnover Number			
3.94	*4.86*	*5.81*	*6.90*	*7.85*	*8.89*	*10.00*
7.8	8.0	7.9	7.9	7.9	7.9	8.0
670	630	655	612	612	595	638
9.7	11.7	9.7	9.9	9.9	10.4	9.7
40.0	46.4	45.5	48.7	48.3	47.6	44.2
5.3	5.1	5.2	4.5	5.2	5.4	5.2
672	490	555	980	999	815	900
54/100	67/100	69/100	66/100	64/100	69/100	65/100
69/31	70/30	72/28	71/29	72/28	69/31	69/31
110	104	119	116	115	115	122
57.1	48.3	56.4	55.5	56.1	54.3	57.8
13.7	17.4	18.9	21.7	15.4	16.7	20.6
88/86	88/87	88/88	90/89	91/90	89/87	88/87
72/78	66/75	67/70	66/73	68/66	66/68	61/63
5.5	6.4	5.7	6.2	6.3	5.9	6.2

8. Resin/crosslinker: determined by infrared analysis of the ester bands of the resin and the triazine bands of the crosslinking agent.

9. Meq/100 grams XC-4010: milliequivalents of amine.

12. Reflectivity: measured with green filter.

12, 13, 14. Film properties determined after baking 20 minutes at 175°C. Substrate was Bonderite 37 (Hooker Chemical Co., Parker Rustproof Div., Detroit, Mich.).

All amine concentrations are expressed as percent neutralization of the acrylic vehicle.
Example:

Steady State (turnover 6-10)
A_1 = unknown
A_2 = 65
A_3 = 20
D = 0.185

Turnover 1
A_1 = unknown
A_2 = 40
A_3 = 20
D = 0.185

The drag out figure has been calculated from data for the first to the tenth turnover. For every one part of paint removed from the bath 0.815 part is deposited and 0.185 part is lost in the rinse water.

$$\text{Steady State } A_1 = \frac{20 - 0.185 \times 65}{1 - 0.185} = 9.8\%$$

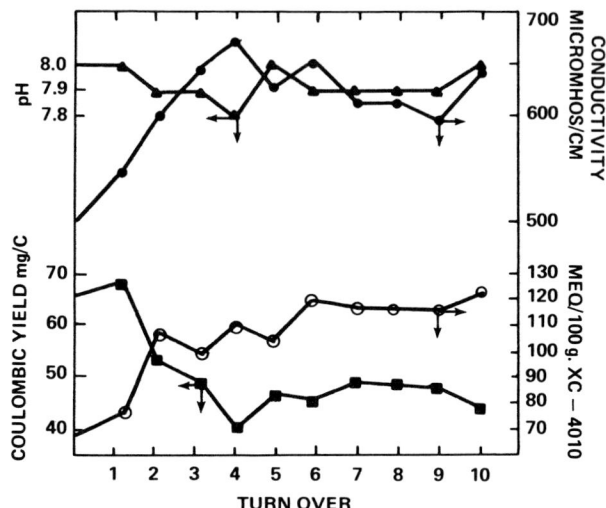

Figure 3. Turnover study of high reflectivity paint. Resin, XC-4010; crosslinking agent, CLA-1116; pigment, Unitane OR-600 titanium dioxide; amine, diisopropanolamine.

$$\text{Turnover 1} \quad A_1 = \frac{20 - 0.185 \times 40}{1 - 0.185} = 15.4\%$$

Thus, the neutralization of the deposited film from turnover 6 to 10 at 65% neutralization of the bath has to be 9.8% for steady state of the bath. For the 40% neutralized bath we would require a 15.4% neutralized deposited film to achieve steady state.

Using the data obtained by the radioactive tracer study (Figures 1 and 2) we find that at a resin neutralization of 60% and a deposition voltage of about 135 volts, 15% of the amine present in the bath is codeposited. Assuming that 15% of the amine is also codeposited at 65% neutralization, we find that at the steady state the deposited film has a neutralization of 9.75%. This number is in excellent agreement with the figure of 9.8% obtained from the turnover analysis.

Applying the radioactive tracer work for 40% neutralization (turnover 1) we find only 6.8% amine codeposited. Comparing the actual codeposited amine of 6.8% with the required amine level for steady state, the amine level has to increase, as we can see from the actual turnover work.

FILM AND DEPOSITION PROPERTIES. The high level of pigment and crosslinking agent in combination with the low level of neutralization in the initial bath give a very high coulombic yield of 68.3 mg/coulomb. The subsequent increase in amine level causes a decrease in coulombic yield. Between the fifth and tenth turnover the coulombic yield fluctuates between about 45 and 50 mg/coulomb. This level is still high compared with most electrocoating systems. The optical and mechanical properties of the baked film remain unchanged from the start of the turnover to the tenth. During the turnover the throwing power decreased because of a decrease in bath age as fresh material was added. With most electrocoating systems a new bath exhibits poorer throwing power than an aged bath. Before making up samples for tenth turnover, the bath was aged for few days prior to testing, and the initial throwing power returned (Table II). Since in practice the rate of turnover is likely to be about once a month rather than once a day, as in this accelerated study, the observed decrease in throwing power probably would not be found in a commercial operation.

Discussion

Analysis of the deposited film gives valuable information regarding the composition of the feed material needed to maintain a steady state. Using the analysis figures, it is possible to predict accurately the feed composition and any disproportionation which might take place in actual operation of a tank. It has been demonstrated that the amine balance in a turnover and the maximum amine level obtained can be determined by using a radioactive tracer method.

The turnover and the radioactive tracer method show excellent agreement with respect to amine content of the deposited film. This agreement signifies that the resin must be uniform in structure and that a minimum of disproportionation could have taken place during the turnover. The analysis of the deposited film in combination with this radioactive tracer method could with most resins reduce or eliminate most of the laborious turnover study work.

Acknowledgment

I thank T. F. Ziegler for performing the radioactive tracer work and D. T. Szokolay for carrying out the turnover study on the depletor. I also thank R. Saxon for his assistance in preparing this paper and for his many helpful suggestions.

Literature Cited

1. Blank, W. J., Koral, J. N., Petropoulos, J. C., *J. Paint Technol.* (1970) **42** (550).

2. Koral, J. N., Blank, W. J., Falzone, J. P., *J. Paint Technol.* (1968) **40,** 156.
3. Bray, G. A., *Anal. Biochem.* (1960) **1,** 279.
4. Brewer, G. E., Strosberg, G. G., Horsch, Martha E., "Preprints," *Am. Chem. Soc. Div. Org. Coatings Plastics Chem.* (1967) **27,** 225.

RECEIVED May 28, 1971.

16

Influence of Solvents on the Electrodeposition of Paint

IVAN H. TSOU and WALTER STUECKEN

Grow Chemical Corp., 14100 Stansbury Ave., Detroit, Mich. 48227

> *The influence of solvents in electrodeposition paint has been studied to determine their effects on polymer viscosity, pH, milliequivalents, current consumption, coating voltage, rupture voltage, and throw power. Two types of solvents were investigated: water miscible and water immiscible. Results show the influence of solvent type on the properties of electrodeposition systems, especially coulombs per gram, throw power, and rupture voltage.*

Solvents have always been a major component of almost all types of paint. The wise selection of the proper solvents has enhanced the success of paint industries in many areas. Early in the development of electrodeposition paint, it was thought that the use of solvents could be avoided. As development progressed, however, it became apparent that solvents are not only necessary but continue to play an important role in the manufacture of electrodeposition paint.

In an electrodeposited paint solvents contribute the following:

(1) In the dispersed phase of an electrodeposited paint, the solvent deposits with the paint film. It helps the flow and the leveling of the film during the bake and hence imparts gloss and eliminates film imperfections, such as pin holes, craters, etc. It also allows the film to coalesce and to form a more impervious film which may improve its corrosion resistance.

(2) In the dispersing phase, or the water phase, the presence of solvents changes the polarity between the paint particles and the water phase and thereby enhances the bath stability.

(3) As a reaction medium during the synthesis of the vehicle resin, solvents help to build up more uniform molecular weight which in turn increases the throwing power and again improves corrosion resistance.

(4) As a dispersing medium, solvent aids the wetting of pigments to the film-forming vehicle resin. Often, it minimizes the disproportionation among the deposition of the various pigments which are present in the paint system.

In several instances during our study we removed the solvent from the paint and found that solvent-free paint exhibited some interesting properties. Generally, it has very high throwing power and requires the lowest power consumption which is indicated by the low values of coulombs per gram of deposited film and the low coating amperage requirement. However, the task of solvent removal is very difficult and costly. We felt the study of the influence of various solvents in electrodeposited paint and the extent of their influence might be of some value. It may help us to select a solvent for an electrodeposited paint.

Since the automotive industry pioneered the development of electrodeposited paint and uses large quantities of it, we selected maleinized oil as the polymer for investigation. The solvents we selected are commonly used in electrodeposited paint supplied to the auto industry—*i.e.*, butyl Cellosolve, isopropyl alcohol and high boiling aromatics. We have added odorless mineral spirits, methyl amyl alcohol, and diisobutyl ketone to include an aliphatic solvent, a water-immiscible alcohol, and an oxygenated active solvent. All three are exempted solvents.

Table I. Viscosity No. 4 Ford Solvent, cp/sec

	Solvent Level of Weight Based on Polymer								
	10	15	20	25	30	35	40	45	50
Butyl Cellosolve	531	228	137	78	—	34	27	—	—
Methyl amyl alcohol	796	—	210	—	93	64	44	33	28
Odorless mineral spirits	—	—	—	—	455	295	209	160	132
Diisobutyl ketone	—	395	168	67	47	29	21	17	15
High boiling aromatics	667	—	305	—	95	63	38	25	21
Isopropyl alcohol	—	212	110	77	46	43	37	—	—

The solvating ability of the solvents was studied. Table I shows that butyl Cellosolve, isopropyl alcohol, diisobutyl ketone, and methyl amyl alcohol reduce polymer viscosity dramatically. Large amounts of odorless mineral spirits are needed to reduce polymer viscosity appreciably.

Table II. Solvent Levels Investigated

% Wt Based on Polymer	10.0	15.0	20.0	30.0
% Wt Based on Bath	0.6	1.0	1.3	2.0

Table II shows the solvent levels that were evaluated. The solvents were blended with the polymer prior to emulsification. All baths were processed to 7.0% total solids. The only composition variable was the solvent level. A control was run using the same bath composition without any solvent. The amine used was diethylamine. Table III shows the parameters investigated and BI method series established by Ford Motor Co., Manufacturing Development Center for evaluating electrodeposition paints. In all testing, their methods were adhered to as closely as possible.

Table III. Parameters and Methods

Parameters	Ford Motor Co., Manufacturing Development Method
pH	BI-20-6
Meq	BI-20-6
Specific resistance	—
Coulombs per gram	BI-20-3
Throw power	BI-20-2B
Rupture	—

Test Results

The results in Table IV indicate that odorless mineral spirits had very little, if any, effect on pH, with the remaining solvents increasing pH by 0.4–0.8 unit. The results also indicate that increasing amounts of solvent did not affect pH except butyl Cellosolve. A Sargent model PBL pH meter was used, and the electrodes were cleaned and standardized prior to determining pH on each sample.

Table IV. pH Values

	Solvent Level, %				
	0	10	15	20	30
Butyl Cellosolve	7.7	8.1	8.5	8.4	8.4
Methyl amyl alcohol	7.7	8.2	8.4	8.4	8.2
Odorless mineral spirits	7.7	7.8	7.8	7.7	7.7
Diisobutyl ketone	7.7	8.0	8.1	8.0	8.0
High boiling aromatics	7.7	8.0	8.1	8.0	8.1
Isopropyl alcohol	7.7	8.3	8.4	8.3	8.4

All the solvents have a decided effect on meq as seen from Table V. As would be expected, butyl Cellosolve and isopropyl alcohol affect meq the most, both being polar and miscible in water. Odorless mineral spirits affects meq the least. Also, the results indicate an increase in meq as solvent levels increase.

As seen from Table VI, specific resistances indicate the same trend as meq, with butyl Cellosolve and isopropyl affecting specific resistance the most and odorless mineral spirits the least. A Yellow Spring Instrument Co. model 31 conductivity bridge with a cell built to specifications from Ford Motor Manufacturing Development Center was used.

Table V. Meq Values (Milliequivalents of Base per 100 ml Sample)

	Solvent Level, %				
	0	10	15	20	30
Butyl Cellosolve	72	105	119	121	128
Methyl amyl alcohol	72	110	96	108	114
Odorless mineral spirits	72	90	87	93	88
Diisobutyl ketone	72	97	115	106	116
High boiling aromatics	72	102	89	106	110
Isopropyl alcohol	72	118	118	129	132

Table VI. Specific Resistance

	Solvent Level, %				
	0	10	15	20	30
Butyl Cellosolve	1375	1080	1040	890	800
Methyl amyl alcohol	1375	880	790	850	910
Odorless mineral spirits	1375	1200	1190	1300	1250
Diisobutyl ketone	1375	1180	930	960	910
High boiling aromatics	1375	1060	950	1000	1060
Isopropyl alcohol	1375	960	910	880	820

The remaining parameters indicate marked effects of solvents on electrodeposition systems. Table VII shows that solvents have a definite effect on coulombs per gram and voltage for 0.7 mil. As for previous parameters, odorless mineral spirits has the least effect, and butyl Cellosolve and isopropyl alcohol affect coulombs the most. Diisobutyl ketone has a marked effect on voltage.

Table VII. Coulombs per gram/volts, 0.7 Mil

	Solvent Level, %				
	0	10	15	20	30
Butyl Cellosolve	69/140	82/90	93/80	96/75	107/60
Methyl amyl alcohol	69/140	89/80	85/80	83/80	85/75
Odorless mineral spirits	69/140	74/100	74/90	66/80	66/80
Diisobutyl ketone	69/140	82/60	81/45	70/40	69/30
High boiling aromatics	69/140	82/80	79/70	78/70	76/65
Isopropyl alcohol	69/140	94/100	112/85	102/80	110/70

Table VIII shows another effect of the solvents. The voltages used to determine throw power were the same as voltages for coulombs per gram and 0.7 mil. The effects of odorless mineral spirits, butyl Cellosolve, and isopropyl alcohol are comparable with other parameters.

Table VIII. Throw Power in Inches

	Solvent Level, %				
	0	10	15	20	30
Butyl Cellosolve	6¼	4	3½	3½	3¼
Methyl amyl alcohol	6¼	4¼	3¾	3¾	4
Odorless mineral spirits	6¼	5¼	4¾	4¾	4½
Diisobutyl ketone	6¼	4	3½	2¾	3
High boiling aromatics	6¼	3¾	3¼	3¼	3½
Isopropyl alcohol	6¼	3	3	2½	2¼

Table IX corroborates results of coulombs per gram and throw power.

Table IX. Rupture Voltage

	Solvent Level, %				
	0	10	15	20	30
Butyl Cellosolve	175	105	95	90	85
Methyl amyl alcohol	175	95	95	90	90
Odorless mineral spirits	175	130	110	95	90
Diisobutyl ketone	175	75	55	55	45
High boiling aromatics	175	95	90	85	80
Isopropyl alcohol	175	120	100	95	85

During our evaluation, other effects of the solvents were noted:

(1) Panel appearance on aged baths. 30% baths were aged under agitation, and panels were coated after 7 days. Appearance of all solvated baths was good while control had craters and pinholes.

(2) Ease of remixing. The 15% solvated baths and control were aged 7 days without agitation. All baths settled, but when placed under agitation, all solvated baths dispersed easily while control required high speed agitation.

The results indicate that all the solvents studied affect the electrodeposition system, especially coulombs per gram, throw power, and rupture voltage. As pointed out previously, the solvents offer advantages for use in electrodeposition. It is up to the formulating chemist to select the optimum solvent or solvent blend needed with the polymer for ease in handling, grinding, emulsification, and stability with the least effect on electrodeposition properties. A blend of odorless mineral spirits with a solvent like butyl Cellosolve or isopropyl alcohol might be optimum, incorporating desirable properties of both. In addition, the low voltage of the diisobutyl ketone system would also merit investigation because of the potential low cost of electrical equipment, unless high throwing power is required. The type of polymer selected is also important. This study should give better insight into the effect of solvents on electrocoat systems, and hopefully it will spur further study and reports in this area.

17

Improved Corrosion Protection through the Electrocoated Edge Spot-Weld Hem Design

GEORGE E. F. BREWER[1] and JAMES W. MITCHELL

Ford Motor Co., Manufacturing Development Center,
24500 Glendale Ave., Detroit, Mich. 48239.

> *It has been verified that slit-shaped rather than circular openings aid in the formation of electrodeposited films on inner surfaces of box-like structures. An improved spot-welded hem construction has been designed and developed recently, which creates slit-like gaps. The results on automobile doors are given as an example for the improved appearance, greater mechanical strength, and much improved corrosion protection resulting from the cooperation of designers, metal workers, and finishers.*

The rapid, worldwide acceptance of the electrocoating process is the result of three main advantages: better corrosion protection, lower cost, and virtual absence of pollution. The increased corrosion protection is mainly the result of the ability to extend paint films into highly recessed areas, such as box sections, channels, and other structures with entrance openings of small open area, leading to relatively large inner areas. Obviously, a larger entrance area will provide easier access for the electric current and will result in a more uniform paint film deposited from any given paint. Usually, however, it is not possible for the engineer and the stylist to provide larger access holes since the material strength and the appearance of the structure must be maintained.

The influence of the size and shape of an opening on the thickness of the electrodeposited paint film, which is found inside of a cavity, has been studied recently (*1*). The authors concluded that for any given electrodepositable paint and structural configuration, the corrosion protection increases with both the open area and the length of the perimeter of the opening. Let us consider a round hole of 0.5 inch id. Its open area

[1] Present address: 11065 East Grand River Rd., Brighton, Mich. 48116.

is 0.20 sq inch, and its perimeter is 1.8 inches. Area times perimeter is then 0.36 cu inch. Now let us consider another open area of 0.20 sq inch, rectangular in shape, 1 inch long by 0.10 inch wide. Its perimeter is 4.20 inches. Area times perimeter is 0.84 cu inch, and a higher degree of corrosion protection is predicted for a structure which carries this type of opening. The problem was the selection of a design which would combine slit-like openings with material strength, and our attention was drawn to welded joints.

Two sheet metal parts, or skins, which form a three-dimensional structure, such as an automobile door, are often designed to be held together by spot-welded hem flanges. In the conventional design of these hem flanges one skin is spot welded into the 180° fold of the other skin (Figure 1, upper left). During the operations of closing and spot welding there is a tendency to distort the outer skin. Experimentations with improved flange designs were therefore indicated, particularly to take advantage of electrocoating and its ability to extend corrosion protective films through slit-like openings.

Edge Spot-Welded Hems

The Ford Manufacturing Development Center, in conjunction with the Ford Metal Stamping Division, developed a new hem weld design in which one skin is bent about 135° to meet an approximately 90° bend of the outer skin (Figure 1, upper right). The spot welds create narrow, slit-like openings between the edges of the two skins (Figure 1, center).

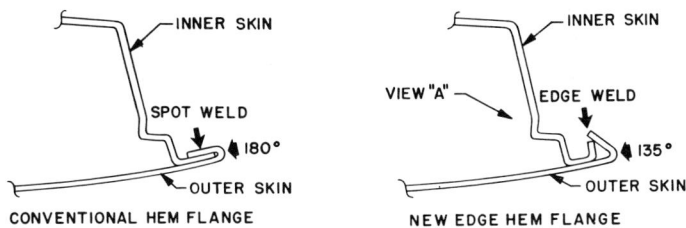

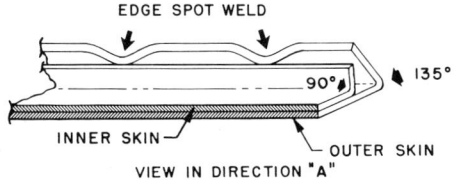

Figure 1. Hem flange spot-welding comparison

These slits aid the electrodeposition of corrosion preventive films on the inner surfaces of the structure.

Automobile Doors. The strength of the new 135° spot-welded hem compares favorably with that of the 180° hem weld. Acceptable welds with adequate strength can be made in varying flange heights. The appearance of the weld is good, and destructive testing pulls "buttons" from the skin. The 135° weld design leaves no marks on the outside metal surface, even when welds are made under widely varying conditions.

A number of automobile doors were built by using the new welded hems in certain areas (Figure 2). These doors were painted by the electrocoat method, then subjected to a variety of tests which demonstrated marked improvements in corrosion protection for inner surfaces.

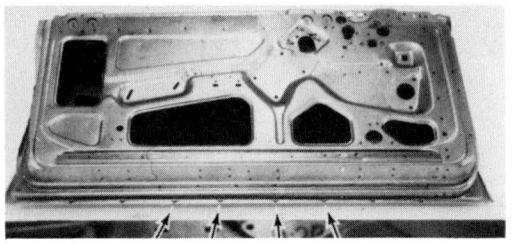

Figure 2. Door with edge spot-welded hem

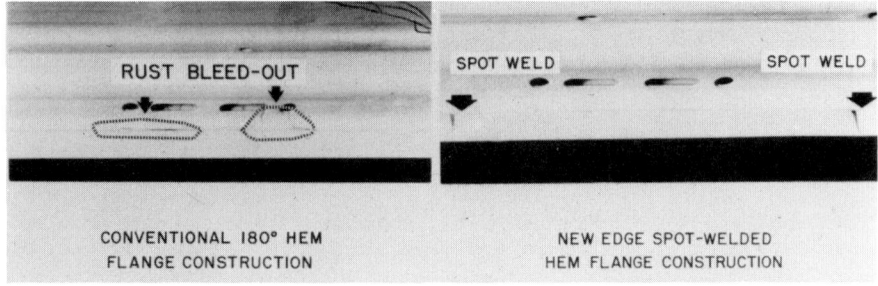

Figure 3. Electrocoated doors after 250 hours of salt spray exposure

Corrosion Protection. Figure 3 compares two electrocoat primed and acrylic enameled doors after 250 hours of salt spray exposure. The bottom edge of one door is formed by a conventional 180° hem flange. It shows slight rust bleeding from the inside and along the hem edge. The other door carries the new 135° hem flange at its bottom edge, and is free from rust after 250 hours of salt spray exposure. Figure 4 shows the test doors after their hem flanges have been cut open. The inner surfaces of the 180° hem show slight corrosion while the inner surfaces

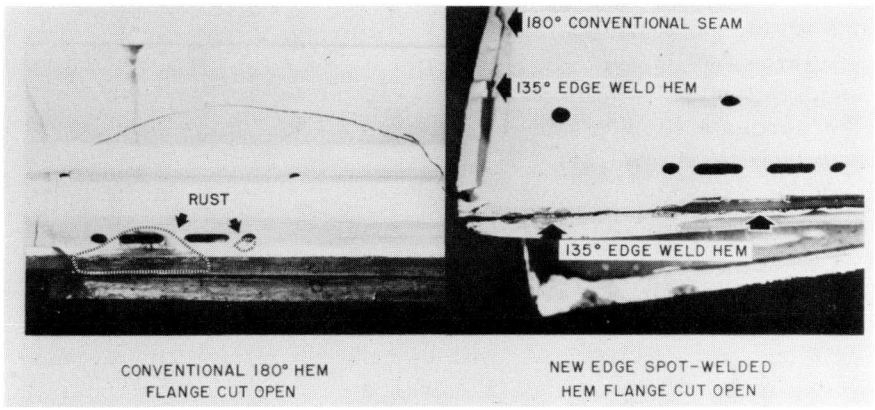

Figure 4. Electrocoated doors after 250 hours of salt spray exposure—hem flanges cut open

of the new 135° hem show a complete paint coat and absence of corrosion.

Summary

The edge spot-welded hem is currently used on a variety of automobile components. It improves the appearance of the final product in several ways since it is free from distortion of the outer surface, results in greater strength, constitutes a minor metal savings, and, in combination with the electrocoating process, results in a higher degree of corrosion protection.

Literature Cited

1. Brewer, G. E. F., Wiedmayer, L. W., "Thickness of Electrocoats Inside of Cavities *vs.* Size and Shape of Openings," *J. Paint Technol.* (1970) **42** (550), 588-591.

INDEX

INDEX

A

Ac input	62
Acrylic copolymers	132
Acrylic polymer	217
Acrylonitrile, copolymer of vinylidene chloride and	101
Adhesion	102
wet	40
wet film	35
Alkali, attack by	81
Alkali cleaner, formulation of	17
Alkaline detergent resistance test	22
Alkalinity	108
Aluminum	
conversion coatings for	10
electrophoretic deposits	102
one-coat systems on	36
pretreatment of zinc and	46
Amine	
codeposition	220
content	218
salts	115
Amines, controlling the concentration of	212
Amino crosslinked electrocoating paints, turnover studies on	216
Amino crosslinking agent	216
Amino resins	113
Ampere-hour meter	67
Analog digital computer	186
Analysis of electrodeposits	132
Analysis of films by infrared spectroscopy	83
Anaphoresis	101
Anion, carboxylate	111
Anodic dissolution	137
Anodic electrode reactions	131
Anodic electrodeposition	111
Anodic polarization	47, 60
curves	50
Anti-corrosion pigments	106
Ash–binder ratio	109
Asphalt emulsions	115
Attack by alkali	81
Automobile doors, electrocoated	234

B

Baking temperatures	54
Balance, material	207
Bath	
agitation and filtration	3
analysis	218
electrocoating	82

Bath *(Continued)*	
maintenance	204
resistance	181
solids level	105
Bonderite 37 pretreatment	83
Boundary layer	183
Butadiene copolymer, styrene–	101
Butyl Cellosolve	228

C

Calcium modified, zinc phosphate process	26
Calcium–zinc process	42
Calculating the concentration of a component	208
Carbides, inorganic refractory	102
Carboxyl containing copolymers	130
Carboxylate anion	111
Cathodic electrodeposition	110
Charge destruction	125
Chromate post treatment	27
Chromate rinse	43
Chromic–chromate rinses	30
Circuit protection	66
Cleaner, formulation of alkali	17
Cleaner–coaters	8
Cleaners	12
Cleaning	15, 40
Coagulation	124
Coated panels	82
Coatings	
aluminum	10
conversion	7, 18
electrodeposited	7, 28
iron phosphate	23, 40
loss and corrosion resistance	19
power supplies for electrodeposition of	62
pretreatment of metals prior to electrophoretic	38
for steel, conversion	9
zinc	9
phosphate	10, 20, 42
Colloids	103
Component addition and removal	208
Component, calculating the concentration of	208
Computer, analog digital	186
Computer calculations	200
Computer simulation	178
digital	191
Concentration, equilibrium	212, 214

Concentration of amines, controlling the 212
Conductivity 108
Controller 63
Conversion
  coatings 18
    for aluminum 10
    for steel 9
    for zinc 9
    and electrodeposited coatings .. 7
Cooling equipment 4
Copolymer
  acrylic 132
  carboxyl containing 130
  styrene–butadiene ....
  styrene–maleic anhydride ... 80
  vinylidene chloride–acrylonitrile 101
Copper, painting metallic 42
Corrosion 17
  filiform 24
  protection 234
  resistance 43, 85
    coating loss and 19
Coulomb/gram ratio 75
Coulombic efficiency 105, 149, 161
Coulombic yield 84, 87, 109
Counterion fixation 130, 140
Counterions 128
Coupling solvent 82
Crosslinking agent 217
  amino 216
Crystal growth 41
Crystal structure, phosphate 53
Crystallization, water of 34
Cured film
  properties 83
  thickness of 109
Curing temperature 86
Current
  consumption 5
  flow as a function of time 169
  measurements, electrocoating .. 70
  required 69
  –time curves 179

D

Dc overload 68
Decarboxylation of fatty acid systems 94
Degree of hydration 44
Deionized water rinse 27, 29, 43
Deposited film composition 84
Deposition
  characteristics, influence of pigment on 104
  effects of temperature on the .. 173
  electrochemistry of polymer ... 149
  wet-on-wet 193
Designing a power supply 76
Detergent resistance 26, 83, 85
Dielectric constant 102
Diels-Alder addition 88
Diethylamine 228

Differential thermal analysis 44
Digital computer simulation 191
Diisobutyl ketone 228
Diisopropanolamine 217
Dispersions 103
Dissolution, anodic 137
Doors, electrocoated automobile .. 234
Drying 30
  phosphate coatings prior to electropainting 44
Dryoffs 27
Dynamic simulation of the electrodeposition of polymers 178

E

Edge spot-welded hems 233
Effect of dryoff 33
Effects of temperature on the deposition 173
Efficiency, coulombic 105, 149, 161
Electrical efficiency vs. voltage .. 197
Electrical layout for step-up voltage process 5
Electrochemistry of polymer deposition 149
Electrocoat
  installations 99
  paints, typical formulation of .. 107
Electrocoated edge spot-weld hem design 232
Electrocoating 232
  advantages of 99
  baths 82
  current measurements 70
  paints, turnover studies on amino crosslinked 216
  pigmentation of 98
  rectifier 63
    control 64
  tank 2
  vehicles, pigmented 102
Electrode reactions 142
Electrodepositable paints 2
Electrodeposited coatings 28
  conversion and 7
  metal concentration found in ... 49
Electrodeposition 1, 4, 47, 191
  anodic 111
  cathodic 110
  of coatings, power supplies for .. 62
  kinetics of polymer 166
  latex 118
  mechanism of 142
  of paint 166
    influence of solvents on 227
  phosphated surfaces after 59
  of polymers, dynamic simulation of the 178
  resins 88, 132
Electrodeposits, analysis of 132
Electrokinetic phenomena 102
Electrolyte drag 28
Electron probe scans 22

INDEX 241

Electroosmosis 104, 135
Electrophoresis 104
  industrial applications of 101
Electrophoretic deposition 1
  pretreatment of metals prior to .. 38
Electrophoretic deposits, aluminum 102
Electrophoretic migration 1
Emulsions 102
  asphalt 115
Ene reaction 88
Epoxy ester 149, 164
  dispersions 114
Equilibrium concentration 212, 214
Esters, epoxy 149, 164
Extender pigments 106

F

Faraday's law 192, 194
Filiform corrosion 24
  resistance test 22
Film-former, electrodeposition of .. 129
Film
  gloss of 105
  growth of 152
  properties, cured 83
  properties, paint 181
  resistivity 149
  rupture 166, 171
  temperature 174
  thickness 149
  unpigmented 82
Filtration, bath agitation and 3
Finite difference approach 186
Fixation, counterion 130, 140
Formulation
  of alkali cleaner 17
  of electrocoat paints, typical ... 108
  variables, pigment 108
Free radical chain polymerization 81

G

Gloss of film 105
Gradient, potential 184
Growth of film 149, 152

H

Heat dissipation 202
Heating, paraffin 56
Helmholtz equation 103
Hems, edge spot-welded 233
High–low voltage control 67
Humidity resistance test 22
Hydration, degree of 44

I

Immersion process 8
Industrial applications of electrophoreses 101
Infrared spectroscopy, analysis of films by 83

Inorganic refractory carbides and oxides 102
Input ac requirements 68
Ions, transport of the 152
Iron oxide pigments 105
Iron phosphate coating 23, 40
Isopropyl alcohol 228

K

Kinetics 166
  of polymer electrodeposition .. 166
Kolbe oxidation 47

L

Lactone 88
Latex electrodeposition 118
Linoleic acid 90

M

Macroions, electrodeposition of ... 128
Macroradicals 146
Maleic anhydride, copolymer of styrene and 80
Maleinization 88
  products 92
  of SOFA 93
Malenizied oil 228
Manual/automatic control 65
Material balance 207
Mechanism of electrodeposition .. 142
Melamine 217
Metal
  concentration in electrodeposited coatings 49
  powders 102
  pretreatment 38, 48
Meters, ampere-hour 67
Methyl amyl alcohol 228
Mica flakes 102
Migration 87
  rate 86
Mineral spirits 229
Minimum deposition voltage 199
Molecular weight studies 144
Morphology, surface 175

N

Neutralization 217
Non-ohmic film resistance 149
Non-volatile content 108
No-rinse process 206

O

Operation of a rectifier 66
Ohm's law 172, 192, 194
One-coat systems
  on aluminum 36
  on steel 36
Oscilloscope, use of 71

P

Paint
  film properties 181
  pH of 39, 108
  solvent-free 228
Painted steel 21
Painting metallic copper 42
Painting, rinsing after 46
Paints
  conventional 27
  electrodepositable 2
  turnover studies on amino cross-
    linked electrocoating 216
  typical formulation of electrocoat 107
Paraffin heating 56
Particle size 104
Passivation acidulated rinse 43
Pen recorder 71
pH 39, 108
Phase growth 132
Phosphate coatings prior to electro-
    painting, drying of 44
Phosphate crystal structure 53
Phosphate stripping 39
Phosphated surfaces 48
  after electrodeposition 59
Phosphating 42
Phosphoric acid deruster 41
Pigment
  anti-corrosion 106
  –binder ratio 105, 219
  on deposition characteristics,
    influence of 104
  extender 106
  formulation variables 108
  iron oxide 105
  polymers and 78
  suspending 106
Pigmented electrocoating vehicles 102
Plating 128
Plexiglas throwing power tank ... 195
Polarization, anodic 47
Polarography experiments 197
Poly(vinyl acetate) 101
Poly(vinyl chloride) 101
Polyelectrolytes, sulfonium 113
Polyester coating 159
Polymer
  acrylic 217
  deposition, electrochemistry of .. 149
  electrodeposition 166
Polymerization, free radical chain 81
Polymers
  and pigments 78
  dynamic simulation of the elec-
    trodeposition of 178
Post treatments 27, 30
Potential gradient 184

Output control 67
Output dc requirements 69
Output voltage 69
Oxides, inorganic refractory 102

Power supply
  designing of 76
  for electrodeposition of coatings 62
  ripple 69
Pretreatment
  Bonderite 37 83
  of metals prior to electrophoretic
    coating 38
  of steel surfaces 42
  of zinc and aluminum 46
Process design 214
Propasol P 82
Properties of SMA-I 82

Q

Quaternary ammonium latex 122

R

Radioactive tracer work 216
Reactions, anodic electrode 131
Reactions, electrode 142
Rectifier
  basic components of 62
  control, electrocoating 64
  electrocoating 63
  operation of 66
Resins
  amino 113
  electrodeposition 88, 132
  synthesis of SMA 81
Resistance 162
  bath 181
  corrosion 85, 106
  detergent 85
  salt fog 106
Rinse 5
  chromate 43
  chromic–chromate 30
  deionized water 27, 43
  passivation acidulated 43
  water 17
Rinsing after painting 46
Ripple filtering 69
Risetime, surge 73
Rupture, film 171
Rupture voltage 40, 105, 231

S

Salt fog
  corrosion resistance 31
  test 22, 83
Salt spray resistance 20, 44
Sand's equation 151
Saponification number 92
SCR's 64
Settling 104
Slit-like openings 233
SMA resins, synthesis of 81
SMA-I, properties of 82
Soaking 54
SOFA, maleinization of 93
Sols 102

Solubilized resins	78
Solubilizer-satisfied and deficient feeds	205
Solvent	
coupling	82
-free paint	228
on the electrodeposition of paint, influence of	227
Specific resistance	105, 230
Spirodilactone	88, 94
Spot-welded hem construction, improved	232
Spray phosphatizing unit	15
Spray process	8
Steel	
conversion coatings for	9
one-coat systems on	36
painted	21
surfaces, pretreating	42
surfaces, zinc phosphated	47
Styrene–butadiene copolymer	101
Styrene–maleic anhydride copolymer	80
Sulfonium	
latex	121
polyelectrolytes	113
salts	110
Surface morphology	175
Surge risetime	73
Suspending pigments	106
Synthesis of SMA resins	81

T

Tank	
electrocoating	2
Plexiglas throwing power	195
Temperature	
baking	54
curing	86
on the deposition, effects of	173
film	174
Testing methods	108
Thermogravimetric analysis	44
Throw power	84, 86, 105, 108, 129, 191, 231
application of	193
definition of	192
vs. material properties	191

Throw power (Continued)	
method of displaying	196
tank (Plexiglas)	195
Time curves, current–	179
Titanium phosphate	41
activator	16
Transformer	65
Transport of the ions	152
Turnover determination	219
Turnover studies on amino cross-linked electrocoating paints	216

U

Unpigmented films	82
Use of an oscilloscope	71

V

Vinylidene chloride and acrylonitrile copolymer	101
Voltage	
applied	104
control, high–low	67
vs. electrical efficiency	197
minimum deposition	199
output	69
rupture	40, 105

W

Water of crystallization	34
Water rinses	17
Wet film adhesion	35, 40
Wet on wet deposition	193

Z

Zeta potential	104
Zinc	
and aluminum, pretreatment of	46
conversion coatings for	9
phosphate	40
conversion coatings	10, 20, 42
process, calcium modified	49
process, nickel and fluoride modified	49
Zinc phosphated steel surfaces	47

QD
1
A355
#119

NOV 8 1974